40.00

Which Contract?

Which Contract?

Choosing the appropriate building contract

Third Edition

Stanley Cox and Hugh Clamp

RIBA Enterprises

© 1989 Stanley Cox and Hugh Clamp

ST. HELENS COLLEGE
690.107 COX
109657
1/2/05
LIBRARY

Third edition 2003

Published by
RIBA Enterprises Ltd
1-3 Dufferin Street
London EC1Y 8NA

ISBN 1 85946 130 1
Product code: 31796

First edition 1989
Second edition 1999

The rights of Stanley Cox and Hugh Clamp to be identified as the authors of this Work have been asserted in accordance with the Copyright, Designs and Patents Act 1988.
All rights reserved. No part of this publication may be reproduced, stored in a retrieval system, or transmitted, in any form or by any means, electronic, mechanical, photocopying, recording or otherwise, without the prior permission of the copyright owner.

British Library Cataloguing in Publications Data.
A catalogue record for this book is available from the British Library.

Publisher: Steven Cross
Editor: Elizabeth Davison
Commissioning Editor: Matthew Thompson
Project Editor: Anna Walters
Design and typography: Ben Millbank
Printed and bound by Hobbs the Printers, Hampshire

This is a completely revised third edition of the book, which appeared first in 1989, and which was later extended by supplements in 1990 and 1996. The second edition was published in 1999. Reference to forms of building contract are to those editions of the forms believed to be current at the time of writing. While every effort has been made to check the accuracy of the information given in this book, readers should always make their own checks. Neither the Author nor the Publisher accepts any responsibility for mis-statements made in it or misunderstandings arising from it.

Contents

Contents continued

Preface

The question 'which contract should we use?' arises frequently at CPD seminars, and despite developments in procurement which sometimes place the client in a more central role, is a matter which still commands the serious attention of architects and other professionals associated with the building industry.

Architects should have an opinion on and be knowledgeable about the most appropriate forms of building contract for their projects. There is little guidance, however, as to what criteria should apply when attempting to make a choice, and decisions are sometimes made without an objective assessment.

This book is written primarily with the interests of architects in mind, but it is hoped that the contents will be useful to others involved on the construction scene. It is accepted that architects today practise in many different ways; from those in consortia engaged on large-scale commissions necessitating sophisticated procedures, through to many who are performing a vital professional role acting as sole consultant on relatively small-scale projects. A majority will be using standard forms of contract, and often the choice rests with the lead consultant.

Forms suitable for use with, say, a new multi-million pound shopping mall, or the restoration of a fire-damaged church, or builders' work in connection with a prefabricated community hall, will obviously all be different. Criteria for choice will include the nature of the project, the procurement method, scope of the work, and the likely complexity of operational procedures. Even so there might be a number of options apparently suitable, still leaving unanswered the question of which particular form will be most appropriate in the given circumstances.

This book does not attempt to provide an authoritative commentary on each of the standard forms referred to. Rather it takes headings commonly found in JCT and other contracts and sets out the main features of each form under these headings, to allow comparisons to be made. All forms of contract are subject to periodic revision, but the information given here is believed to be accurate at the time of writing. With each form referred to, the comparative summary is followed by subjective comment. It is hoped that this will be of positive help to all those facing the sometimes difficult question 'Which Contract?'.

The authors would like to thank Matthew Thompson and Katy Banyard of RIBA Enterprises, together with all those users and publishers who willingly supplied helpful guidance and information. They particularly acknowledge the help of the Scottish Building Contract Committee for assistance with contracts adapted for use under Scots law.

Stanley Cox Hugh Clamp

Introduction

As long ago as 1870 the RIBA, with the London Builders' Society, issued an agreed document called Heads of Conditions of Builders' Contract, which outlined the principles to be applied when a building contract was being drafted. In 1895 the RIBA issued its own Conditions of Contract, revising them in 1902 to allow, for the first time, alternatives for contracts with or without quantities.

A further revision of the contract, agreed by both organisations, was issued in 1909, and remained available until the Joint Contracts Tribunal was established in 1931 and produced its own version. The 1931 form was revised in 1939 and was later adapted for use by local authorities. It remained current until well after the Second World War.

In 1963 the Joint Contracts Tribunal published a new Standard Form of Building Contract with two versions, private and local authority, both with alternative editions for use with or without bills of quantities.

In the years that followed, changes in the building industry were extensive and dramatic. Developments in construction technology required different building techniques and patterns of working, which in turn generated the need for alternative methods of contracting and financing. There was corresponding pressure for contract documentation to follow suit. The Joint Contracts Tribunal was expanded to reflect a wider range of interests and thereby a more effective consensus for its agreed standard forms. The 1980 edition of the Standard Form, which replaced JCT63, allowed a total of six variants. Each of the two versions (private and local authority) was available for use with quantities, without quantities and with approximate quantities. These were constantly updated and made more adaptable through option clauses, supplements and a succession of Amendments, until the advent of a consolidated 1998 Edition of the Standard Form of Building Contract.

In 1998 the Joint Contracts Tribunal became a body with a distinct legal persona – The Joint Contracts Tribunal Limited – and a registered business address. Its present concern is to upgrade those documents previously issued by the Joint Contracts Tribunal, and to work towards what the Latham Report (see below) referred to as 'a complete family of interlocking documents'. In recent years, other interested professional bodies have published their own contract forms in response to the increasing variety of contractual and building requirements. Such bodies responsible for forms in general use today include:

- The Stationery Office;
- Institution of Civil Engineers;
- Architecture and Surveying Institute (formerly the Faculty of Architects and Surveyors);

- Association of Consultant Architects;
- Scottish Building Contract Committee;
- Joint Council for Landscape Industries.

Many of these bodies publish more than one standard form. The result of all this well-intentioned publishing activity in the building industry is that today's clients, and their advisers, are faced with a bewildering choice of forms.

The Latham Report (*Constructing the Team*, by Sir Michael Latham, HMSO (1994)) has had a considerable effect on contract forms and procedures. This was a wide ranging look at the industry as it then existed. It recommended better project strategy, more integrated ways of working, fairer tendering, improved payment procedures, and the rapid resolving of disputes by adjudication. It stressed the place of clients as the driving force, the importance of a full brief, design quality to be considered alongside the lowest price to determine best value, fair dealing for all parties in an atmosphere of mutual cooperation, and the outlawing of unfair conditions. It laid the foundations for new thinking about procurement, the place of partnering, and the benefits of long-term relationships.

The Egan Report (*Rethinking Construction*, the report of a Construction Task Force chaired by Sir John Egan, HMSO (1998)), built on many of the points raised by the Latham Report, and took further the call for greater efficiency. In particular, it stressed the need for improvements in construction costs, construction time, predictability concerning delivery, and a reduction in the number of building defects. It called for the elimination of wastage through lean thinking, and a reduction in the number of site accidents. The idea of targets, performance measurement indicators and benchmarking as aids to achieving planned and consistent improvement, reflected a changing climate of opinion on procurement.

These two reports, together with Part II of the Housing Grants, Construction and Regeneration Act 1996 which followed as a direct result of the Latham Report, have significantly affected procurement methods, and the text of current contract forms.

Which contract?

Many of the forms published by The Joint Contracts Tribunal Limited and the other bodies listed above are featured in Chapters 6 to 13 of this book, under the headings of the respective procurement paths. The preceding chapters examine the interrelation of contracts for professional services and for building, recommend ways to evaluate clients' requirements realistically, and discuss the choice of building procurement, contract type, and contract form. The final chapter of the book sketches out some problem scenarios and suggests possible contractual approaches.

Thinking about contracts 1

Context

The Latham Report recommended modern contracts which provide for:

- fair dealing for all concerned – a team effort;
- wholly related documents, whether for the appointment of consultants, contractors, sub-contractors or suppliers;
- easily comprehensible language backed by guidance notes;
- mechanisms to allow allocation of risk to be adjusted;
- fair payment provisions to agreed procedures and a realistic timescale;
- fair and rapid resolution of disputes, albeit subject to final arbitration.

These points make a useful checklist against which any form of building contract can be measured.

The points have been embraced by most publishers of construction contracts. In some cases they were already close to being attained by modern drafting, in other cases amendments were necessary to bring compliance with the latest legislation, and Latham Report recommendations were incorporated at the same time.

Architects will be primarily concerned with two categories of contract: those under which professional services are provided and those under which construction work is carried out.

Although different in purpose, the two categories are essentially complementary. The authority and obligations of the professional acting for the client during the construction period might be significantly conditioned by the wording of the building contract. Likewise, the part played by the professional during the pre-contract stages, although determined largely by the contract for services, will be influenced by the type of procurement to be adopted and the role to be played by other professionals and by the contractor. To this extent at least, the choice of the one contract might be influenced by the content of the other.

Contracts for professional services

Leaving aside the fact that many architects offer a wide range of skills not always directly buildings specific, it is for this latter role that most are commissioned. Nowadays, architects increasingly find themselves asked to provide initially only a partial service, or a service which is not 'traditional' in the sense that they might not be the lead consultant. Sometimes they start out as the client's agent before eventually becoming part of the contractor's professional team. Some architects may gladly rise to the challenge of an entrepreneurial role as developer, others may be suited to a construction management role. But whatever format they take, architect's services need to be covered by an appropriate contract.

Some forms of appointment may require bespoke drafting because of the nature of the services to be provided, and it is increasingly common to find clients who insist on their own forms or who will require amendments to standard forms. However, in this book it is assumed that any contract for architect's services is likely to be on one of the RIBA standard appointing documents with its Memorandum of Agreement or Letter of Appointment, Conditions of Engagement, Schedule of Services, and Schedule of Fees etc. It is important that the appointing document clearly identifies whether the commission relates to a role either as lead consultant, Contract Administrator, or design leader, and is compatible with the procurement method.

It might be that the appointment can initially only be up to a defined Plan of Work stage, and will need to be extended if the project proceeds. It might be that the Agreement will require amending if the role of the architect changes, or if there are shifts in the procurement procedures. The RIBA standard appointing documents are flexible and can be adapted for use in most situations.

Where the appointment is for services as a sub-consultant, it is vital to check that the Agreement is a back-to-back form as far as terms are concerned, and that the conditions are preferably compatible with those of standard RIBA forms of appointment. Equally if additional forms are proposed, whether these relate to partnering charters or collateral warranties, then a careful check is essential to ensure that the obligations are no more onerous than those which arise from the main appointing document.

Contracts for building

Building construction today often entails complex and intensive site operations, with huge sums of money locked into development programmes. Contractors may have partial or total design responsibilities, and in addition may undertake demanding management and coordinating roles; patterns of working have never been so diverse. This change of emphasis, together with the increasing range and scale of work, has inevitably led to a proliferation of alternative forms of building contract, both standard and purpose drafted.

Using standard forms

For the majority of building contracts under which architects have a stated role, standard printed forms are readily available, and these should be used without alterations. Bespoke amendments can easily impair the balance of the forms and the precise meaning of the contract conditions could be a matter for endless argument between lawyers. There is also the practical point that if a contractor is asked to tender knowing that it is proposed to use a special or amended form of contract, his first action will be to pass it to his legal advisers for checking. The cost of this will then be reflected in his tender. It is sound advice to resist alterations and amendments wherever possible.

The Latham Report recommended that 'All parties in the construction process should be encouraged to use those Standard Forms without amendment'.

Non-standard agreements

However, there may be instances where no standard form of building contract fits the client's requirements and a specially drafted agreement is needed. There is an increasing trend towards non-standard agreements on commercial or larger contracts. Any such drafting should be entrusted to a lawyer with the appropriate specialist knowledge, and he or she should always be engaged directly by the client. Architects without legal training and experience in such matters are strongly advised not to attempt even seemingly minor changes to standard wording or drafting of additional clauses which might make published documents non-standard.

The JCT forms

As noted in the Introduction, several bodies produce building contracts and sub-contracts, but by far the greatest number is published by The Joint Contracts Tribunal Limited (JCT). A whole family of documents is now available to cover a wide variety of work which ranges from jobbing or maintenance and repair work on the one hand to management contracting on the other.

The JCT forms also have the unique distinction that they are produced by a tribunal on which there is representation of the professions, the industry and client bodies. These documents are held to be fair and balanced in the interests of the parties – they are truly 'agreed'. This means that they are unlikely to be interpreted as contra proferentem by the courts, if the meaning of any condition is at issue. It also means that, as the contracting organisations have been represented at the drafting stages, they can tender with more confidence.

It should be noted, however, that at a time when legislation has tended to intervene in the common law of contract, any contract in the construction field is likely to be subject to:

- Unfair Contract Terms Act 1977;
- Sale and Supply of Goods Act 1994;
- Unfair Terms in Consumer Contracts Regulations 1994 (SI 1994/3159);
- Construction (Design and Management) Regulations 1994 (SI 1994/3140) ('CDM Regulations');
- Housing Grants, Construction and Regeneration Act 1996 (Part II in particular);
- Late Payment of Commercial Debts (Interest) Act 1988;
- Contracts (Rights of Third Parties) Act 1999.

Advising the client

Where the architect provides full professional services, it may be his or her duty 'to consult with and advise the Employer as to the form of contract to be used' (*Hudson's Civil Engineering and Building Contracts* 10th Edition, 1970). These words might not find favour with the majority of architects today, and it has been said that the best course is to give the 'for and against' leaving the employer to make the choice. Interestingly, the endorsement on many 1998 JCT forms of contract refers to the client engaging 'a professional consultant to advise on and to administer its terms'. Thankfully the JCT has produced a most helpful Practice Note 5 Deciding on the Appropriate JCT Form of Main Contract, which is clearly written with the interests of the client in mind.

It is thought that so far there has been no case recorded of an architect being successfully sued for recommending a contract which was later held not to have been appropriate. But the range of available forms is increasing, and Amendments (particularly in the case of JCT forms of contract) need to be published to keep them current and consistent. Just keeping up to date with such changes is often a taxing business. Also, there is sometimes a disconcerting time lag between the publication of authorised Amendments and instructions confirming their use, particularly by bodies in the public sector who have a strong influence upon such matters. Advising on the edition of a form to be used and whether particular Amendments are to be incorporated extends the architect's need for care in carrying out this duty and requires a thorough knowledge of the state of the art.

Advising on forms of contract is unlikely to be easy in the wider European context. Problems can arise in construing the meaning of provisions which may have been poorly translated, or which may be subject to interpretation under a different legal system than that which obtains in the courts of England and Wales. Problems can also arise when working on projects overseas, where the contractor or suppliers are based in another country, and perhaps the law of the contract has not been clearly established. Many standard forms can be adapted for use under the laws of Northern Ireland or Scotland but not all. Some specifically state that they are not for use in Scotland. The Scottish Building Contracts Committee has published its own range of forms specifically for use under Scots law.

Recommending a contract for a particular set of circumstances is not something to be undertaken lightly: it requires a knowledgeable and methodical approach. The contract in question should first be studied carefully both in respect of the content and conditions, and also for its appropriateness in the known circumstances.

Thinking about contracts

1

Watchpoints

- Give thorough consideration to contract matters and choice at the earliest possible time.
- Remember that contracts for professional services and building contracts, although distinct, may have implications for each other. The method of procurement has repercussions on roles and documentation.
- Assuming that the procurement method has not been decided before the architect is appointed as lead consultant, it may be the architect's duty to advise on the appropriate contract. If so, do not allow this to be left to other consultants.
- Ensure that contract agreements and conditions are set down in writing and the agreement executed before work starts.
- Recommend the use of standard forms of contract whenever possible. Bear in mind that JCT forms have the added merit that they are 'agreed' forms.
- If particular circumstances require specially drafted or extensively amended contracts, it is essential that legal advice is taken.
- Before recommending a form of contract, study its contents and understand its implications. It is important to set up a control system which ensures that stipulated procedures are followed.

There are still a few architects who minimise the significance of contract wording by referring to the importance of trust. They blandly talk of the best kind of contract as being the one which is put into the drawer and only taken out when things go wrong. An appropriate contract, fairly and firmly administered, is the best means of trying to ensure that things do not go wrong in the first place!

Despite the Egan Report's stated ambition to replace contracts with performance measurement and bring reliance on formal documents to an end, for most present situations the cautious advice must remain – never proceed on the basis of a 'handshake' agreement. Whilst it cannot be denied that an atmosphere of trust and confidence is desirable, it is total folly to rely on some 'understanding' that has not been fully documented and properly executed. Different people can have genuine but quite different understandings of what was agreed.

As architect Ronald Green once pointed out in *The Architect's Guide to Running a Job*, Architectural Press (1986): 'The difficulty about a gentleman's agreement is that it depends on the continued existence of the gentlemen'!

Establishing a contract profile 2

There may be clients who choose to misinterpret the Egan Report's message of increased efficiency and think that it is now possible to construct a quality building at breakneck speed for a knock-down price. Any such unfounded euphoria needs to be dispelled at the outset.

The reality is that although the three most important considerations for any client are usually cost, time and quality, the business of building procurement invariably calls for some compromise or conscious balancing of these priorities. This requires adequate thinking time and careful thought.

We live at a time when visual presentation for such management tools as benchmarking, performance indicators and the like has become part of life. To some all this might just be peripheral jargon, but for an increasing number of construction professionals so-called 'radar charts' are a useful reminder of key points to be considered and performances recorded. Therefore it might prove an interesting exercise to set up a kind of radar chart in advance of any proposed building contract. This could be simply a manual exercise to help focus the mind on the balance of priorities – for example what level of quality is required, how much time is available before construction and for operations on site, and to what extent and in what respects cost considerations are paramount.

The profile that emerges might also suggest where design responsibilities are to rest and their extent, and the most suitable procurement methods and construction procedures. This in turn will affect tendering arrangements and the amount and format of the information needed.

Figure 2.1 is a radar chart which presents the three elements of cost, time, and quality in terms of contract priorities. Even where they are not in conflict, these elements need to be reconciled and ideally balanced. Asking the right questions at the right time might result in a visual profile which allows quick comparisons, and help in arriving at the appropriate contract.

Figure 2.1 indicates the contract priorities for a project at the minimum capital cost to be built in the shortest possible time. There must be reasonable certainty over cost and timing, and all this combined with the desire for reasonable design quality from the client's consultants!

Taking these three elements in greater detail:

2 Establishing a contract profile

Figure 2·1: A contract profile

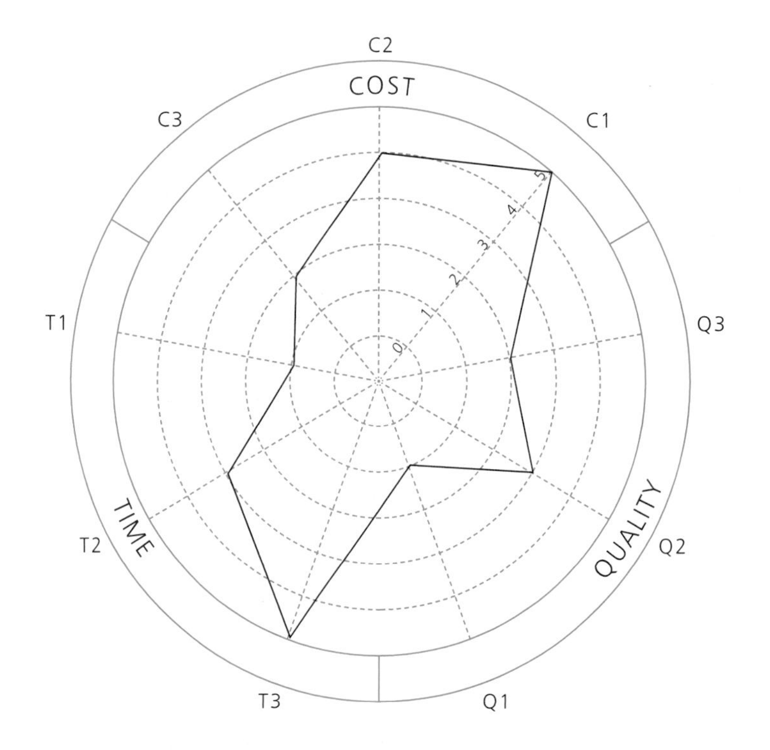

	Criteria		Priority (0 lowest–5 highest scale) 1	2	3	4	5
COST	C1	Lowest possible capital expenditure					✓
	C2	Certainty over contract price, no fluctuation				✓	
	C3	Best value for money overall		✓			
TIME	T1	Earliest possible start on site	✓				
	T2	Certainty over contract duration			✓		
	T3	Shortest possible contract period					✓
QUALITY	Q1	Top quality, minimum maintenance	✓				
	Q2	Sensitive design, control by employer			✓		
	Q3	Detailed design not critical, leave to contractor		✓			

2 Establishing a contract profile

Cost

In this country, at least until quite recently, cost has been the decisive factor in building contracts. The repercussions are well known. Design briefs have often been subject to unrealistic cost constraints from the outset, sometimes even failing to distinguish between initial capital outlay and long-term costs. The quantity surveyor's role has assumed greater importance because of the need for stringent financial control throughout the project. To many client bodies, lump sum contracts have seemed attractive because they promise, at least in theory, cost certainty. Design and build or management contracts have been let on the basis of a guaranteed maximum lump sum, although the design information is known to be incomplete at the time of tendering. Despite the considerable risks which sometimes attend keen competition, all too frequently the lowest price has been the determining factor.

Now the emphasis appears to be shifting and although the Egan Report called for greater efficiency resulting in lower prices, the best value for money factor is now being encouraged.

It is particularly important to clarify some of the cost considerations, by addressing such questions to the client at the outset. For example the lead consultant might wish to raise:

Q. *Is there a set limit, which the Contract Sum must not under any circumstance exceed?*
Comment: the client might have limited funding and require the reassurance of a fixed price with no risk of fluctuation.

Q. *Is it essential to know precisely the cost of the work before operations start on site?*
Comment: this suggests that the work needs to be measured and described in detail at tender stage – possibly by use of firm quantities.

Q. *Once the contract is signed, in the event of variations is the client authorised to incur additional expenditure?*
Comment: it may be that the client has to rely heavily on some 'once and for all' external source of grant aid or other funding, and there may be conditions attached.

Q. *Is the contract to be awarded to the lowest tenderer regardless of other considerations?*
Comment: value for money will not always be achieved by this process. Designing down to a figure might mean excessive running and maintenance costs later on, which the client should take into account. There might also be stipulations concerning quality, imposed by a grant aiding body where this is conservation work.

Q. *Is the tenderer expected to allow for any increases in labour and materials etc. when pricing?*
Comment: it may be in the interests of the client to forgo cost certainty and accept

a price based on known factors, assuming that fair and controlled increases can be dealt with under fluctuations provisions. Much will depend on the state of the market at the time, on the anticipated duration of the job, and on the projected means of recovering the capital expenditure.

Q. *Is total accountability an imperative? Must every penny be satisfactorily accounted for?*
Comment: public bodies in particular are subject to the scrutiny of the auditor, and are also understandably sensitive about any allegations of dubious practice.

Q. *Does the client have rules or standing orders which require evidence of competition when sub-contractors or suppliers are invited to tender?*
Comment: commercially minded clients often need it to be demonstrated that no opportunity for the best deal has been overlooked. Many public bodies have approved lists for selected items.

Q. *What is understood by 'cost considerations'? Capital expenditure only, or are maintenance and life cycle costs also to be taken into account?*
Comment: the lead consultant has a duty to advise the client of any long-term implications.

Quality

No one really expects to buy a Rolls Royce for the price of a family saloon, yet there are clients who conveniently overlook the fact that in this world you only get what you pay for. It is essential to agree what is meant by 'quality' both in respect of services and the finished project, and to define what measurable standards are to apply. For example the lead consultant might wish to raise questions such as:

Q. *Are considerations of commercial prestige and public image likely to influence the degree of quality desirable?*
Comment: some client bodies have a house style which expresses the efficiency or market-commanding confidence of the company, and which they would wish to see embodied in their buildings.

Q. *Is it important to use high quality materials?*
Comment: for example, the project might be located in an environmentally sensitive spot such as a conservation area, or perhaps subject to stringent planning permission conditions. On the other hand, the project might just be an envelope for some retailing or industrial process.

Q. *Is this a listed or historic building where alterations or extensions will require higher than average standards of craftsmanship?*
Comment: it may not be possible to establish how much work is likely to be entailed, or to describe and measure it before opening up. Yet control by the client as work proceeds is essential, and any body responsible for funding might impose conditions.

Establishing a contract profile

2

Q. Is it essential that all matters of design and specification are firmly under the control of the design team, and that specialist sub-contractors or suppliers are nominated or otherwise selected by the client?

Comment: where the architect is lead designer, in the absence of anything to the contrary, this is likely to bring at first instance design responsibility. Even where design work is to be undertaken by others, the responsibility for coordination and integration into the overall design will remain with the lead designer.

Q. Is it safe to entrust certain design details to the contractor's own organisation (with agreement to sub-let) without the quality of the design concept overall being impaired?

Comment: unless a contractor's design obligation is expressly referred to in the building contract, this is unlikely to be implied under standard forms of contract. Even where the contractor or sub-contractor is responsible for some design input, it might be essential to make a provision for designs to be submitted to the client before work is actually carried out.

Q. Is the building intended to be relatively maintenance-free?

Comment: this is not an attainable goal! The best that can be achieved is minimum maintenance. So what is the expected life span? When will major components be in need of upgrading or replacing? Is the overall appearance intended to last for a fashionable cycle only and due for a face-lift which will not involve structural change? Is there need for such eventualities to be programmed at the outset?

Q. Does it make good commercial sense for the client to provide for a constant site presence to monitor the contractor's control of quality?

Comment: should the contract allow for this, and if so by what means will it be achieved? What might be expected of the contractor in terms of quality management and perhaps evidence through relevant KPI information?

Q. Can some or all of the work only be performed by a firm with specialist expertise?

Comment: this might mean selecting a contractor after interview and negotiation rather than by competitive tendering. It might result in appointing the contractor at design stage in order to benefit from advice on construction methods and materials, and possibly to provide specialist design input. It would probably affect the production of information. It could result in appointing a specialist firm as the main contractor, leaving general builders' work to be sub-contracted.

Time

Quality and speed are not obviously happy bedfellows. Where fast construction relies on the use of numerous prefabricated major components and systems, much will depend on whether manufacture has taken place under monitored and controlled conditions, in order that specified performance can be guaranteed.

2 Establishing a contract profile

Good supply-chain management can reduce both wastage and construction time. Savings can also be achieved through overlapping detailed design and construction stages, where the contractor controls the flow of necessary information, and where the contract allows the contractor freedom in the choice of sub-contractors and substitutions. Real savings in overall project time comes through effective management, and not by taking short cuts like premature rushed starts.

Clients, who sometimes take a long time to make up their minds to proceed, are understandably disappointed when an immediate start on site is not advisable. Whatever method of procurement is adopted, it is important to allow sufficient time for all the relevant matters to be properly considered at the pre-contract stage. Design and build, for example, depends largely for its success on whether the client's requirements have been carefully worked out, and time should be allowed for this. The overall time of a project from inception to completion is of greater significance than the time taken for site operations alone. Having said all that, the construction industry has a poor record of excessive construction times, and failure to complete on time. Predictability is an important factor for most clients.

Time considerations can significantly affect the profile of the contract and must be explored thoroughly. For example, the lead consultant might raise with the client questions such as:

Q. *Is there time for a full brief to be systematically developed, so that the client's detailed requirements can be properly reflected in the tender documents?*
Comment: this will be essential for those tendering who need full information when submitting a lump sum price. Anything less could bring the uncertainty of remeasurement. However, in the event of rapid action being required, say in the case of restoring damaged premises, a hand-to-mouth approach might be unavoidable.

Q. *Does the client have heavy rental or other financial commitments which mean that the earliest date for completion is likely to be an overriding objective?*
Comment: as above this might be an overriding consideration which justifies the earliest possible start, and accepting the probability of some uneconomic working.

Q. *Are there any commercial or other external pressures which make it imperative to complete by a certain date?*
Comment: the need to catch seasonal trade for instance, or to be completed in time for a major event which is already programmed and immovable. In such a case a guaranteed completion date might be needed despite the contractor pricing for the risk.

Q. *Is phased or sectional completion necessary?*
Comment: for example, to allow some office units to be occupied, or parts of an industrial complex to be commissioned, ahead of completion of the whole contract. Not all contracts will allow for this, and to be effective it needs to be included in the contract conditions as a programmed requirement.

Q. Is it desirable to phase possession by the contractor and limit it to successive parts of redevelopment of the works, in order that business can continue during the building work?
Comment: during redevelopment within a site, is it essential to control phasing and also to accommodate some decanting of occupants and processes during site work? Is it necessary to restrict operations to certain times of day or intermittent periods? Many standard building contracts appear to assume exclusive possession by the contractor, and with small domestic projects in particular this is not practical and some allowances may be needed.

3 Which procurement method?

In some instances the procurement method will have already been decided before an architect is appointed, either as the result of company or authority laid down policy, or because circumstances or constraints leave very limited options, or because the choice has already been made by the client advised by a lead consultant who is not the architect.

It is generally accepted that there are three methods of procurement currently practised in this country. The traditional or conventional approach, in which at least in theory design and construction are seen as separate elements; design and build, which implies a more integrated approach; and management, by which either the client or a contractor assumes the central management responsibility. However, there are in addition many variants, hybrids, or compound versions of these methods.

Which procurement method is likely to prove the most appropriate in a given situation will depend upon the nature and scope of the work proposed, how the risks are to be apportioned, how and where responsibility for design is to be placed, how the work is to be coordinated, and on what price basis the contract is to be awarded.

An important point to remember is that the choice of form of contract cannot usually be settled until the procurement method and the type of contract have been established. This will mean considering the following:

Design responsibility

Design has been defined as devising an arrangement, then specifying the components needed to realise that arrangement, and lastly detailing a method of joining or erecting those components. Design can mean the overall concept or form of the building, it can relate to the component parts including specialist installations, or can be the result of meeting specified criteria for durability, performance etc.

It is important to establish:

- how, if at all, design responsibilities are to be apportioned between the architect as lead designer, other consultants, the contractor and specialist sub-contractors and suppliers;
- what contractual provisions will apply to the design of the works.

Coordination responsibility

This might include responsibility for workmanship, goods and materials, working methods, programming, ordering, general coordination and supervision.

3 Which procurement method?

It is important to establish:

- what contractual arrangements will apply for the coordination of the work;
- whether the works are to be carried out under a single contract, or under a combination of separate contracts, either in sequence or in parallel.

Price basis

A contract might be let on the basis of a lump sum price, or if this is not possible or desired, then measurement to some agreed basis might be the only practical option. Alternatively a cost-plus approach might be appropriate, although there is the risk that the final figure could differ greatly from the first estimated cost. The questions of what tendering methods are most suitable, and what tender documents will be needed, will rest on the choice of procurement method.

Plan of Work

The procurement method to be adopted and type of contract will have implications for the 'plan of work'. This logical division of a project into stages was devised by the RIBA in the 1960s and has since become accepted throughout the building industry. The Outline Plan of Work 1998 moves from pre-design stages (Feasibility) through design and construction (Pre-Construction Period) to post-construction (Construction Period) activities. It has been widely adopted as the basis for calculating consultants' fees, and gives a very useful description of work stages, particularly in traditional methods of procurement.

With Plan of Work the client and the appointed professionals are involved throughout the project, but the design and construction work stages are separated as is usual with traditional procurement. This results in a linear pattern (see Figure 3.1).

With design and build, although similar work stages are still present, they are not so compartmented. The contractor will normally be involved at design stage, to an extent depending on how much he is responsible for scheme design as opposed to developing a design already produced by the client's consultants and embodied in the client's requirements. Some of the work stages are arranged in a different sequence, permitting parallel working or fast tracking to save time overall. The contractor will normally continue detailed design during construction stages (see Figure 3.2).

Plan of Work is still relevant to management procurement, whether management contracting or construction management. Figure 3.3 shows an admittedly over-simplified picture of operations. On a large project there might, for example, be 50 or more works or trade packages and the operational pattern can become very complex. There needs to be considerable involvement and collaboration between the consultants and the managing contractor throughout, as parallel working continues and abortive work can easily occur.

Procurement using traditional methods

In the traditional approach, the client accepts that consultants are appointed for design, cost control, and contract administration, and that the contractor is responsible for carrying out the works. The responsibility of the latter extends to all workmanship and materials, including work by sub-contractors and suppliers. The contractor is usually chosen after competitive tendering on documents giving complete information. However, the contractor can be appointed earlier, either through negotiation or on the basis of partial or notional information.

The traditional method, but using two stage tendering or negotiated tendering, is sometimes referred to as the 'Accelerated Traditional Method'. By this variant, design and construction can run in parallel to a limited extent. Whilst this allows an early start on site, it also entails less certainty about cost.

Watchpoints

- A traditional lump sum approach requires the production of a full set of documents before tenders are invited. Adequate time must be allowed for this.
- The traditional procurement method assumes that design will be by appointed consultants, and it does not generally imply that the contractor has any design obligations. If this is to be the case, for example with specialist sub-contract work or performance specified work, express terms should be included in the contract.
- Because the client appoints consultants to advise on all matters of design, and cost, he thereby retains control over the design and quality required.
- There is certainty of cost, to the extent that a lump sum is known before work begins, even if it has to be adjusted during the construction period as provided for in the contract.
- The contractor depends heavily upon the necessary information and instructions from the architect being issued on time. There is a risk of claims if they are delayed. Information release dates are sometimes agreed beforehand, but making these contractually binding can cause problems.
- The client may decide which specialist firms the contractor is to use, although the contractor will require certain safeguards relating to performance.
- All matters of valuation and payment are the responsibility of the client's consultants.
- If it is impossible to define precisely the quantity or nature of some of the work, it is still possible to adopt a traditional method on the basis of approximate quantities, provisional sums or cost reimbursement. However, this is a less than perfect solution: the fuller and more accurate the information, the nearer to the relative safety of the lump sum approach.

- There are widely accepted codes of procedure for single stage or two stage tendering, whether competitive or on the basis of a negotiated price. These should be used whenever possible.

Figure 3·1: Plan of Work stages: traditional procurement

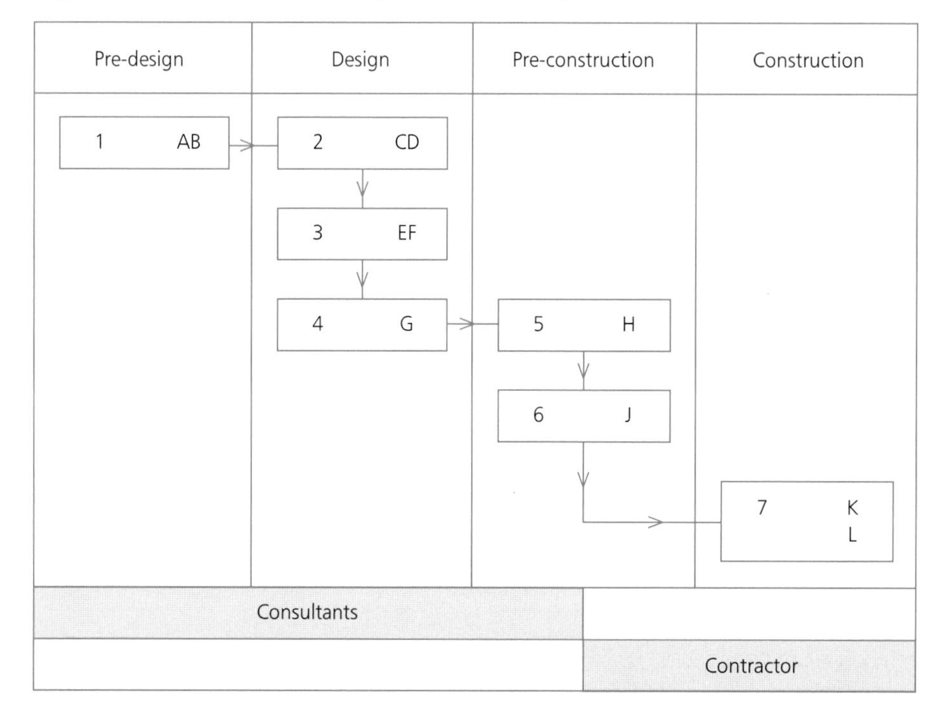

Key
1 Appraisal and strategic briefing work by consultants
2 Outline and detailed proposals by consultants
3 Final proposals and production information by consultants
4 Tender documentation by consultants
5 Tender action – appointment of contractor
6 Mobilisation by contractor
7 Construction to Practical Completion and after completion

Procurement using design and build methods

The client may need to appoint consultants to advise on his design requirements and costs, if he does not have this expertise available in-house. The contractor is responsible to a greater or lesser extent for design, as well as for carrying out the work. The arrangement may be for total design and construction, or for design development and production information based on a scheme design supplied by the client's consultants.

Figure 3·2: Plan of Work stages: design and build procurement

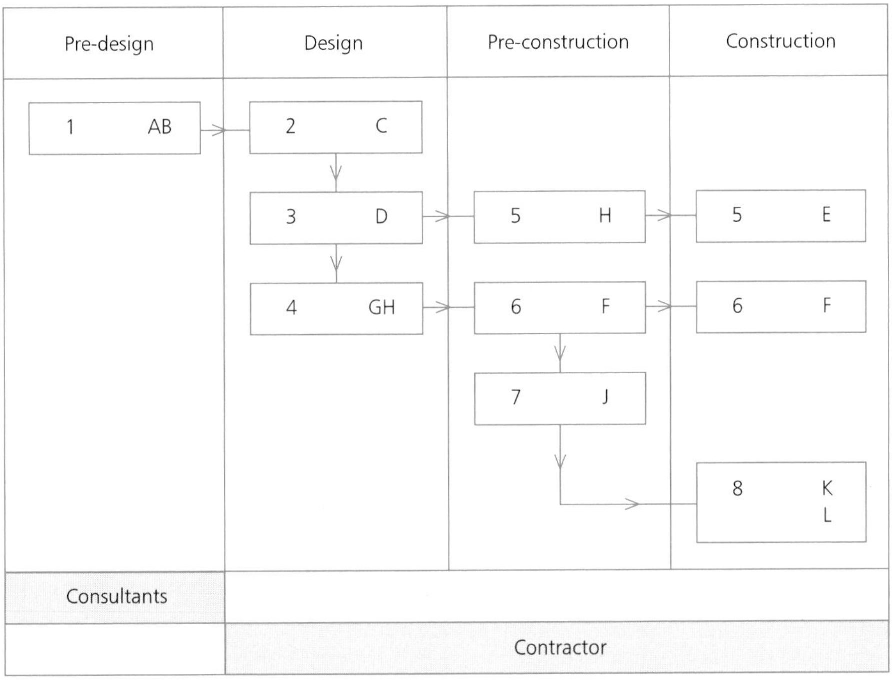

Pre-design	Design	Pre-construction	Construction
1 AB	2 C		
	3 D	5 H	5 E
	4 GH	6 F	6 F
		7 J	
			8 K L
Consultants			
	Contractor		

Key

1 Appraisal and strategic briefing work by consultants
2 Outline proposals/client's requirements
3 Detailed proposals by contractor
4 Tender action including contractor's proposals
5 Final proposals by contractor and specialists
6 Production information
7 Mobilisation by contractor
8 Construction to completion and after completion

The contractor may be appointed either by competitive tender or as the result of a negotiated agreement. Where a design and build agreement is negotiated with just one contractor, it is sometimes referred to as 'Single Direct Design and Build'. Where an approach is made to a number of contractors, even if this is a two stage operation with only the most promising proceeding to the second stage, the agreement is sometimes referred to as 'Competitive Design and Build'. It tends to take slightly longer, but it usually results in a more developed design and greater certainty of cost and timing.

Watchpoints

- In the most straightforward of design and build contracts, in theory there is usually a single point of responsibility. The client therefore has the advantage of only one firm to deal with – and one firm to blame if things go wrong. In practice, however, the client's requirements are often detailed to the extent that the contractor's design contribution, and therefore liability, is diminished.

- The client lacks control over detailed aspects of design; however, this might be acceptable where the broad lines of the scheme are satisfactory and the detail relatively less important.
- Construction work can be started early as a great deal of detailed design work can proceed in parallel. It is mainly the contractor, however, who benefits from the operational flexibility.
- Responsibility for completing on time rests wholly with the contractor. There should be little risk of claims because of allegations that information from the client is late. This obligation on the contractor to be responsible for the flow of his necessary information is one of the most attractive features of design and build.
- There is greater certainty of cost, even to the extent that, if required, responsibility for investigating site and sub-soil conditions can be made entirely the contractor's. Any significant changes in the client's requirements will affect the Contract Sum however, and are likely to prove costly.
- Often the client requires that the contractor appoints the client's consultants to develop the design under a consultant switch agreement or by novation. If this is not the case, it is always advisable to ask for information about who the contractor intends using as designer. Adequate professional indemnity insurance should always be a requirement.
- The client should appoint consultants to advise on the preparation of the requirements, and it is important that adequate time is allowed for these to be properly prepared.
- The requirements might include specific items, or even provisional sums, but generally it is prudent to prescribe performance criteria, so that a high degree of reliance is placed on the contractor.
- In the absence of any stipulations to the contrary, the contractor's design obligations are absolute. However, they are usually reduced in standard forms of contract to those of the professional's duty of using reasonable skill and care.
- Valuation of changes by the client is entirely the responsibility of the contractor, and the client has no quantity surveyor to intervene on his behalf.
- It is often difficult to evaluate design and build tenders objectively where both schemes and prices are submitted. Tenderers should be informed of the criteria to be used, and whether price is likely to be the prime consideration.
- Benefits can arise from designers and estimators having to work closely together. The contractor's awareness of current market conditions and delivery times can ensure

that a contract runs smoothly, economically and expeditiously.

- The client's agent or representative should be selected with great care. He or she can be a key member of the employer's organisation, a professional consultant, a project manager or, depending on the nature of the work, a clerk of works. The extent to which an agent is empowered to act for the client needs to be clearly established.

Procurement using management methods

There are several variants of management procurement practised in this country, but management contracts and construction management are the two most common. With management contracts the client usually starts by appointing consultants to prepare project drawings and a project specification. The management contractor is selected by a process of tender and interviews, and paid on the basis of the scheduled services, prime costs and management fee.

His role is literally to manage the execution of the work, and he is not usually directly involved in carrying out any of the construction work, which will be done in 'packages' undertaken by works contractors usually appointed by the management contractor. In some procurement arrangements, the management contractor might also accept a design liability. In 'management contracting', works contractors are directly and contractually responsible to the management contractor. Its coordinated approach and potential for flexibility results in greater operational speed and efficiency.

Where the management contractor's obligation is total – where, that is, he or she accepts responsibility for both design and construction – the arrangement is sometimes referred to as 'Design, Manage, Construct'. It is usually featured as a contractor-led procurement method, but there is no real reason why it cannot equally well be architect-led. Indeed, where small works are sometimes carried out under direct trades contracts and coordinated by the architect in the absence of a main contractor, this comes close to being such a procurement method. It does, however, demand a degree of highly specialised expertise and experience in setting up and managing operations which most architects are unlikely to possess.

With 'construction management' agreements, which will often be client drafted, there is usually a lead designer responsible for overall design, a construction manager responsible for the management and coordination of work, with the client responsible for directing the project and entering into all trades contracts. As the trades contractors are directly and contractually responsible to the client, the construction manager is in some ways less accountable for time and costs, whilst the client takes on the greater risk.

Watchpoints

- Management procurement methods are best suited to large, complex, fast-moving projects where early completion is desirable.

- This method of procurement depends upon a high degree of confidence and trust. There is unlikely to be a guarantee of firm contract price before the work actually starts on site, and the decision to go ahead usually has to be taken on the basis of an estimate on project information.
- The management contractor acts for the client, and should therefore put the client's interests first throughout the job.
- It is essential to appoint the management contractor at an early stage, so that his knowledge and expertise are available to the design team throughout the crucial pre-construction period.
- Much of the detailed design work can be left to proceed in parallel with the site operations for some work packages, thus reducing the time needed before the project starts on site. Indeed, a great deal of detailed design will need to be left to specialist sub-contractors or suppliers.
- The client has a considerable degree of flexibility on design matters. The design can be adjusted as construction proceeds, without sacrificing cost control. However, coordination of design work for components or elements can be difficult, with an attendant risk of costly abortive work.
- The management contractor can select specialists and order materials on long delivery in good time without any of the uncertainties and complexities which attend traditional nomination procedures.
- Although the project proceeds on the basis of a contract cost plan only, effective cost control is still possible with the help of an independent quantity surveyor.
- A competitive tendering element is retained for all works or trades contracts, which usually account for most of the overall prime cost. Tenders for works packages will normally be on a lump sum basis.
- This method of procurement is most appropriate for large-scale, complex projects, so that only large construction firms with the relevant experience are likely to prove suitable. At present these are relatively few in number. Similarly, construction management is only likely to be of interest to experienced clients with in-house expertise to undertake the high degree of involvement needed.
- The management contractor's staff may lack the necessary experience and have difficulty in adjusting to the idea of abandoning a simple work-for-profit motive in favour of providing a service in the interests of the client. Before recommending the appointment of a management contractor it is essential to interview the key staff involved.
- Above all, a management contractor or construction manager should be appointed because of his or her assumed or preferably proven ability to manage. He or she will

need to use and be familiar with a variety of sophisticated techniques to deal with the coordination of what is often a large number of works contracts.

Figure 3·3: Plan of Work stages: management procurement

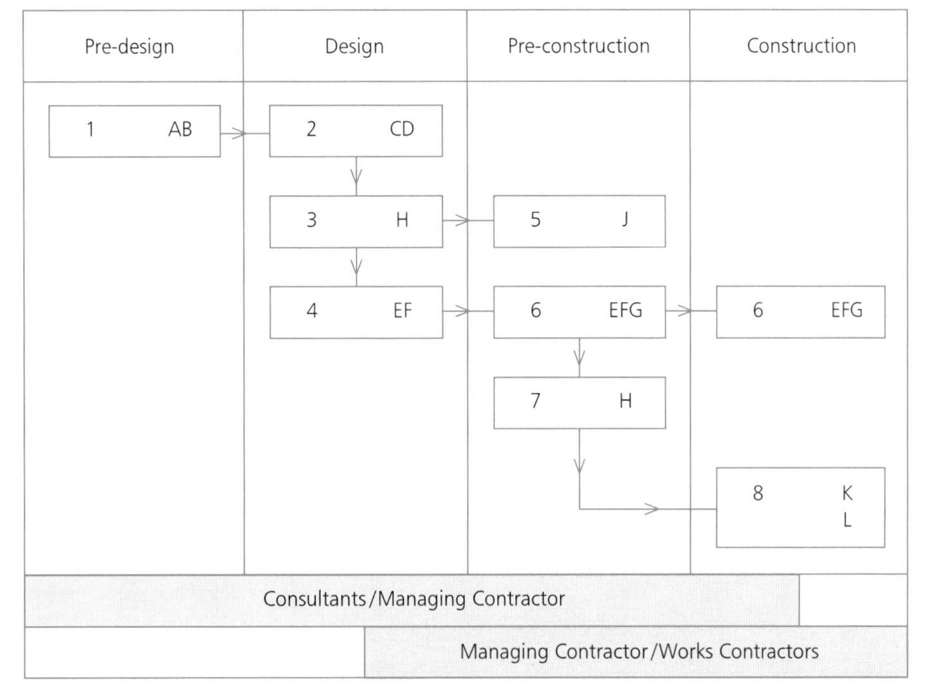

Key

1 Appraisal and strategic briefing by consultants and contractor
2 Outline and detailed proposals
3 Appointment of contractor and agreement on trade or works appointments
4 Final proposals and production information (continuing process)
5 Mobilisation
6 Production information and coordination of works packages
7 Tender action and adjustments (continuing process)
8 Construction to Practical Completion and after completion

Contractual relationships

The pattern of contractual and functional relationships shifts or changes according to the procurement method adopted.

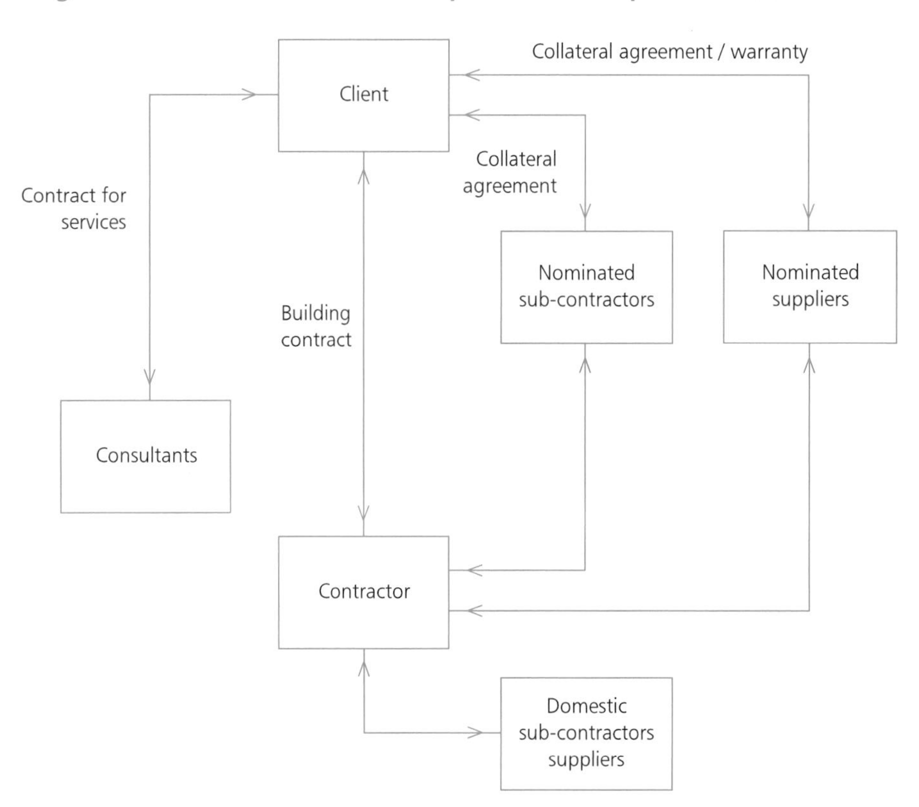

Figure 3·4: Contractual relationships: traditional procurement

In the traditional approach, the client is in direct contractual relationship with the consultants on the one hand and the contractor on the other. Any contractual links for sub-contracts or sales contracts will be between the contractor and the firms in question. Only where the client makes nominations is it advisable to recommend collateral agreements to safeguard his interests in respect of any matters which might lie outside the building contract.

Figure 3·5: Contractual relationships: design and build procurement

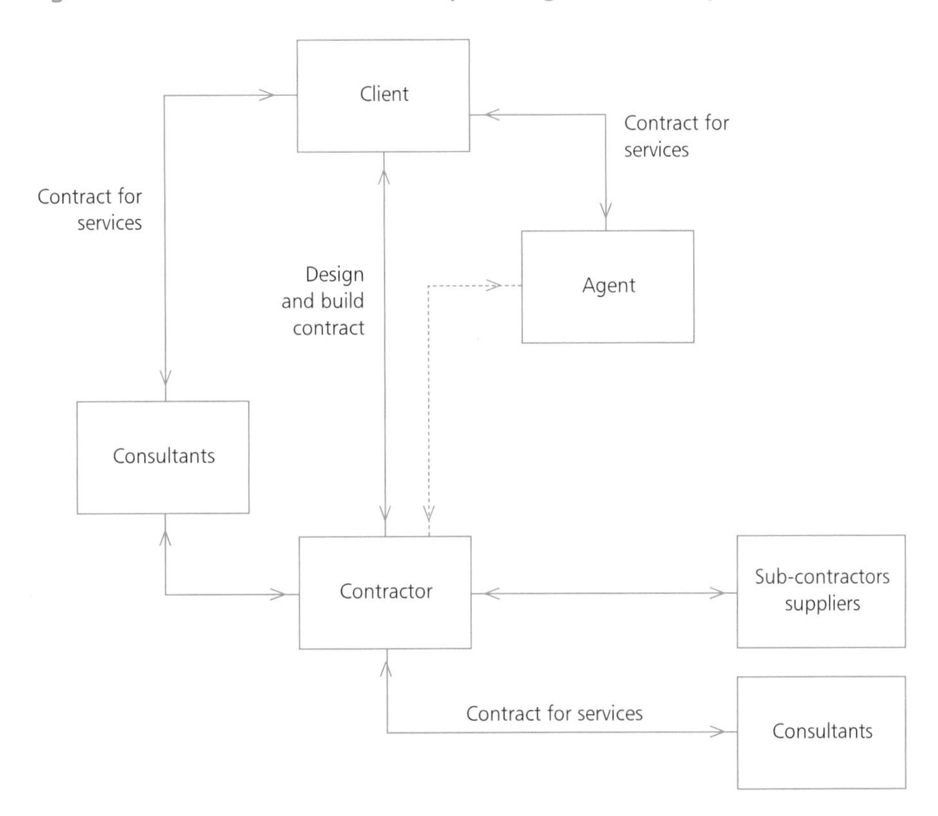

With design and build, it is likely that in the absence of in-house professional staff, the client will wish to engage outside consultants to advise on the preparation of requirements and to evaluate and select tenders etc. The main contractual link is between the client and the contractor and the client's agent or representative has only a limited role. The contractor might also have a contractual link with his own design consultants, and with sub-contractors and suppliers. As the contractor is wholly responsible for their performance, both in terms of design and construction, there might be less need for collateral agreements between them and the client.

Figure 3·6: Contractual relationships: management procurement

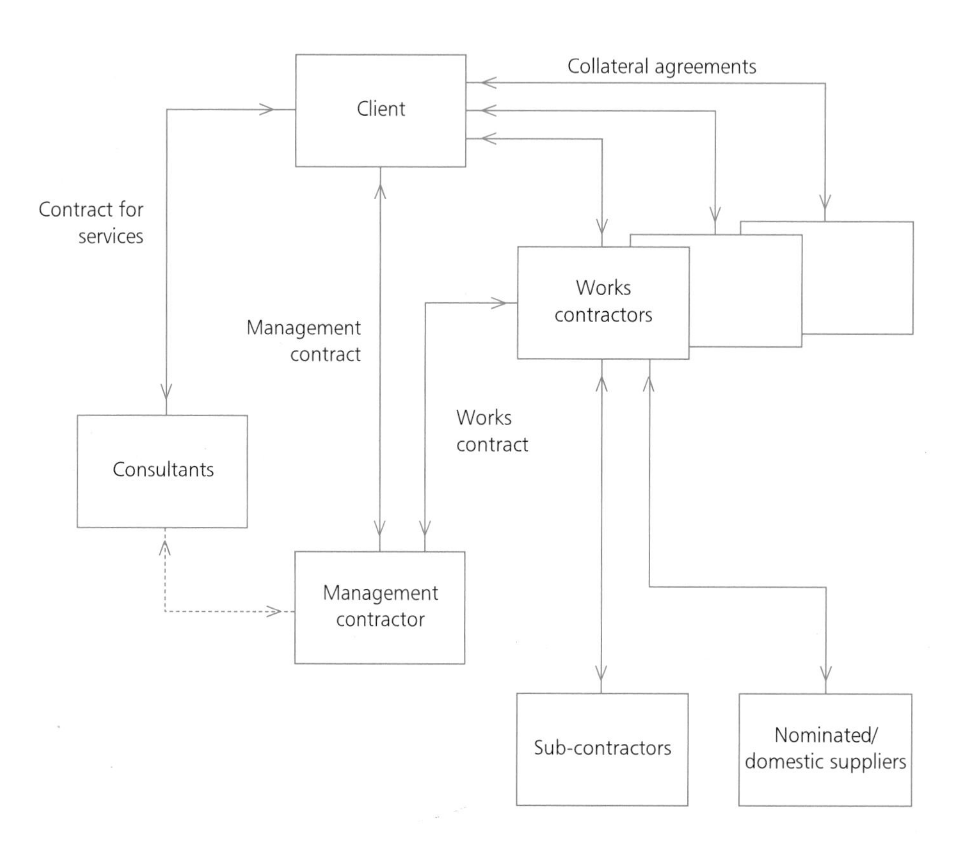

The contractual relationship in a management contract is between the client and the management contractor, with all works contractors in direct relationship with the latter. It may also be desirable to establish a contractual relationship between the client and each works contractor by means of a collateral agreement. In construction management the contractual relationship is between the client and the construction manager, with all trades contractors in direct relationship with the former.

Assessing the risks

In every building contract there is some degree of risk. People may be injured or property damaged. This category of risk, often referred to as pure and particular risk, is usually covered by the appropriate insurance. Contract conditions often make it a contractual obligation to take out the cover required (for example, against injury to persons, or damage to property due to fire, storm, water, collapse, subsidence, vibration, etc.).

Another category of risk is fundamental risk. This will include damage due to war, nuclear pollution, supersonic bangs, etc. Such incidents are all the subject of statutory liability, and no insurance cover is normally available – or needed. The third category, often referred to as speculative risk, is something which can be apportioned in advance as decided by the parties to a contract. This may include losses in time or money which are the result of unexpected ground conditions, exceptionally adverse weather, unforeseeable shortages of labour or materials, and other similar matters wholly beyond the control of the contractor. It is essential to define at the outset who is to bear losses arising from such events.

With traditional lump sum contracts the intention is that there should usually be a fair balance of speculative risk between the parties. The balance can be adjusted as required, but obviously the greater the risk to be assumed by the contractor, the higher the tender figure is likely to be. The apportionment of risk accepted by the parties also varies considerably depending on the type of contract.

As can be see from Figure 3·7, the balance of speculative risk will lie almost wholly with the contractor in the case of a design and build contract, where a complete package is supplied. Conversely, the balance is most onerous for the client where the management procurement path is adopted.

Figure 3·7: Speculative risk

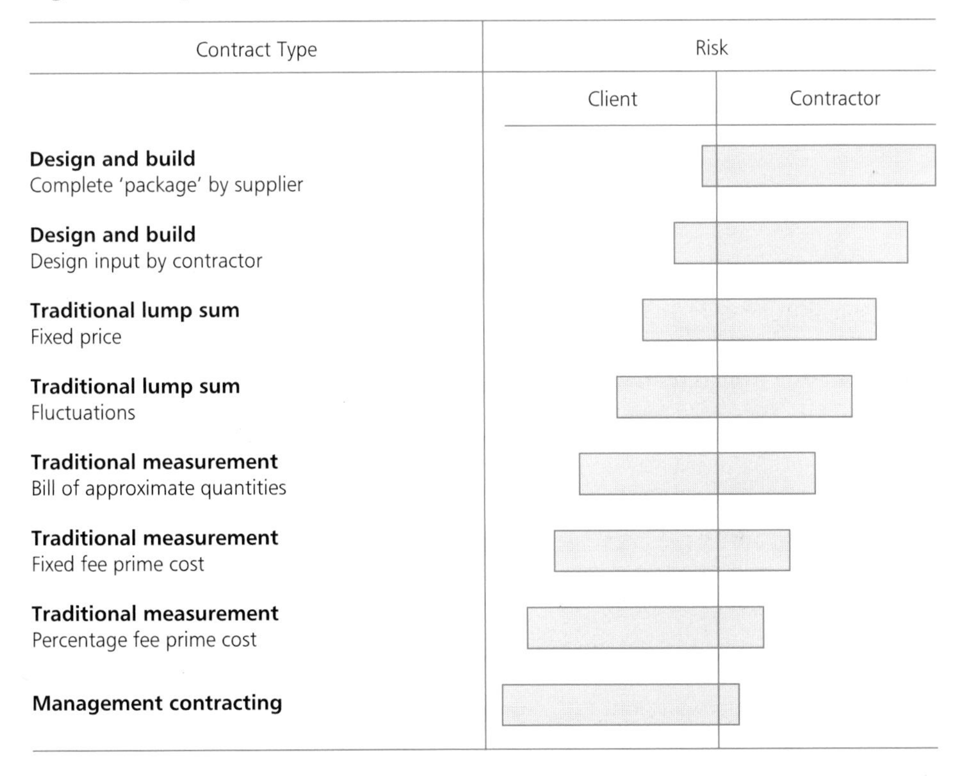

Figure 3.8: Comparison of procurement methods

	Speed	*Complexity*	*Quality*	*Flexibility*
Traditional	Not the fastest of methods. Desirable to have all information at tender stage. Consider two stage or negotiated tendering.	Basically straightforward, but complications can arise if client requires that certain sub-contractors are used.	Client requires certain standards to be shown or described. Contractor is wholly responsible for achieving the stated quality on site.	Client controls design and variations to a large extent.
Design and build	Relatively fast method. Pre-tender time largely depends on the amount of detail in the client's requirements. Construction time reduced because design and building proceed in parallel.	An efficient single contractual arrangement integrating design and construction expertise within one accountable organisation.	Client has no direct control over the contractor's performance. Contractor's design expertise may be limited. Client has little say in the choice of specialist sub-contractors.	Virtually none for the client once the contract is signed, without heavy cost penalties. Flexibility in developing details or making substitutions is to the contractor's advantage.
Management	Early start on site is possible, long before tenders have even been invited for some of the works packages.	Design and construction skills integrated at an early stage. Complex management operation requiring sophisticated techniques.	Client requires certain standards to be shown or described. Managing contractor responsible for quality of work and materials on site.	Client can modify or develop design requirements during construction. Managing contractor can adjust programme and costs.

	Certainty	*Competition*	*Responsibility*	*Risk*	*Summary*
	Certainty in cost and time before commitment to build. Clear accountability and cost monitoring at all stages.	Competitive tenders are possible for all items. Negotiated tenders reduce competitive element.	Can be clear-cut division of design and construction. Confusion possible where there is some design input from contractor or specialist sub-contractors and suppliers.	Generally fair and balanced between the parties.	Benefits in cost and quality but at the expense of time.
	There is a guaranteed cost and completion date.	Difficult for the client to compare proposals which include for both price and design. Direct design and build very difficult to evaluate for competitiveness. No benefit passes to client if contractor seeks greater competitiveness for specialist work and materials.	Can be a clear division, but confused where the client's requirements are detailed as this reduces reliance on the contractor for design or performance. Limited role for the client's representative during construction.	Can lie almost wholly with the contractor.	Benefits in cost and time but at the expense of quality.
	Client is committed to start building on a cost plan, project drawings and Specification only.	Management contractor is appointed because of management expertise rather than because his fee is competitive. However, competition can be retained for the works packages.	Success depends on the management contractor's skills. An element of trust is essential. The professional team must be well coordinated through all the stages.	Lies mainly with the client – almost wholly in the case of construction management.	Benefits in time and quality but at the expense of cost.

Which type of contract? 4

For traditional procurement

Under this method, with design separated from construction, most building contracts are strictly work and material contracts. The basic assumption is that the contractor is to carry out and complete the work as shown on or described in the documents supplied by the client, and is responsible for his own working methods. This might include the design of any temporary works necessary to enable him to achieve the desired result. Apart from this, and in the absence of anything to the contrary expressly in the contract, he has no responsibility for design. The contract wording may of course be extended to expressly include for a limited design obligation by the contractor.

The contractor is responsible for workmanship and materials and this is in respect of complying with the standards of the contract as far as workmanship is concerned and of satisfying implied terms of merchantable quality etc. concerning goods and material etc. This will normally extend to all sub-contractors of whatever description or status. The contractor will also be responsible for any defects, other than those which are directly attributable to faulty design or arise from misuse. He is not normally responsible for maintaining the building: any such requirement would have to be expressly included in the contract.

Basically there are three types of contract available with the traditional procurement method:

- lump sum contracts: where the Contract Sum is determined before construction starts, and the amount is entered in the Agreement;
- measurement contracts: where the Contract Sum is accurately known only on completion, and after remeasurement to some agreed basis;
- cost reimbursement contracts: where the Contract Sum is arrived at on the basis of the actual costs of labour, plant and materials, to which is added a fee to cover the overheads and profit.

Lump sum contracts

The contractor undertakes to carry out a defined amount of work in return for an agreed sum. This can be a fixed amount not subject to recalculation, in which case there would be no opportunity for the client to make variations after work has started on site. The sum is more likely to be subject to limited fluctuations, usually to cover tax etc. changes not foreseeable at the time of tendering. The sum may be subject to fluctuations in the cost of labour, plant and materials – the so-called 'full fluctuations' provisions. Recovery may be by use of a formula, or by the tedious business of checking vouchers, invoices, etc.

Lump sum contracts 'with quantities' are priced on the basis of drawings and a firm bill of quantities. Items which cannot be accurately quantified can be covered by an approximate quantity or a provisional sum, but these should be kept to a minimum.

Lump sum contracts 'without quantities' are priced on the basis of drawings and another document. This may simply be a specification of a descriptive kind, in which case the lump sum will not be itemised, or one that is detailed to the extent that the Contract Sum is the total of the priceable items. The job might be more satisfactorily described by Schedules of Work, where the lump sum is the total of the priced items. In the latter cases, an itemised breakdown of the lump sum will be a useful basis for valuing any additional work. Where only a lump sum is tendered, then a supporting Schedule of Rates or a Contract Sum Analysis will be needed from the tenderer.

Tenders can be prepared on the basis of notional quantities, but they will need to be replaced by firm quantities if it is intended to enter into a 'with quantities' lump sum contract.

Measurement contracts

These are also sometimes referred to as 'remeasurement' contracts. This is where the work which the contractor undertakes to do cannot for some good reason be accurately measured before tendering. The presumption is that it has been substantially designed, and that a reasonably accurate picture of the amount and quality of what is required is given to the tenderer. Probably the most effective measurement contracts, involving least risk to the client, are those based on drawings and approximate quantities.

Measurement contracts can also be based on drawings and a Schedule of Rates or prices prepared by the client for the tenderer to complete. This type of contract might be appropriate where there is not enough time to prepare even approximate quantities, or where the quantity of work is very uncertain.

Obviously the client has to accept the risk involved in starting work with no accurate idea of the total cost, and generally this type of contract is best confined to small jobs.

A variant of this is the measured term contract under which rates can be established for categories of work, although instructions or orders will be required before any single job in the anticipated programme is carried out.

Cost reimbursement contracts

These are sometimes referred to as 'cost plus' contracts. The contractor undertakes to carry out an indeterminate amount of work on the basis that he is paid the prime or actual cost of labour, plant and materials. In addition he receives an agreed fee to cover management, overheads and profit. Checking the prime costs which are directly related to the works is relatively straightforward. The variable is the fee, which should be agreed beforehand, and establishing precisely what it covers. The basis of the fee can give rise to many variants of cost plus contracts. Which is likely to be the most appropriate will depend on the particular circumstances.

Cost plus percentage fee

The fee charged is directly related to the prime cost. It is usually a flat rate percentage, but it can also be on a sliding scale. However, the contractor has no real incentive to work at maximum efficiency, and this variant is only likely to be considered where requirements are particularly indeterminate pre-contract.

Cost plus fixed fee

The fee to be charged is tendered by the contractor. This is appropriate provided that the amount and type of work is largely foreseeable. The contractor has an incentive to work efficiently so as to remain profitable within the agreed fee.

Cost plus fluctuating fee

The fee varies in proportion to the difference between the estimated cost and the actual prime cost. The presumption is that if the latter cost increases due to the contractor's supposed inefficiency, then the fee will be reduced accordingly. This approach depends upon there being a realistic chance of ascertaining the amount and type of work at tender stage.

Cost reimbursement based on a target cost

This is a slight variant on the previous type. The fee is related to an agreed target. The actual prime cost above or below the target affects the fee earned. On the whole this type of contract is not popular, and is likely to prove complex to administer.

It is of course always open to the client to pay direct for the cost of labour, materials, plant, hire charges etc., with the contractor receiving only an agreed fee for managing the execution of the works.

For design and build procurement

This procurement method gained acceptance in the early 1980s with the introduction of a 'with contractor's design' variant of the standard JCT form. Recently it has been endorsed by the public sector as a preferred procurement option, and most publishers of standard forms now produce documents which allow for a contractor's design responsibility.

The contractor may have responsibility for some or all of the design of the project. The contract wording must expressly refer to this, and the extent of the design obligation needs to be set out as clearly as possible.

The requirements of the client can be stated briefly and simply, perhaps comprising little more than a site plan and schedule of accommodation. However, they are more likely to be stated in a document of several hundred pages with precise specifications, accompanied by a well worked out concept design. The design role of the contractor might even be restricted to developing design detail and preparing production information based on the design information supplied by the client.

Unless the contract states otherwise, it seems that the liability for design is an absolute liability under which the contractor warrants fitness for the purpose intended. However, usually standard design and build forms expressly limit the design liability of the contractor to the normal professional duty to exercise reasonable care and skill. Independent consultants engaged by the contractor are therefore under a liability no greater than normal. An indemnity or acceptance of liability is likely to be worthless unless backed by adequate indemnity insurance, and this is something that should be checked before a contractor is appointed. Where the contractor does not have in-house designers and intends to use outside consultants, their identity should be established before a tender is accepted.

Generally it is better to specify in terms of the performance requirement rather than to prescribe in detail, because this leaves the responsibility for design and selection firmly with the contractor. However, with some types of development (housing, for example) the more precise the requirements the less likely it is that a tenderer will start out with the fixed intention of utilising proprietary components or his own standard shell designs.

It is claimed that design and build contracts offer certainty on the Contract Sum and bring cost benefits. The close integration of design and working methods, and the relative freedom of the contractor to use his purchasing power and market knowledge most effectively, might suggest this. Unfortunately, it is often very difficult to find out just how competitive the figures are.

On the matter of speed, it should be possible to ensure a quicker start on site, and the close integration of design and construction should result in more effective programming. However, time is needed for the client's consultants to prepare an adequate set of requirements, and time is also needed to compare and evaluate offers and schemes from competing tenderers. The success or otherwise of a design and build operation depends to a large extent on the client properly setting out the requirements in the first place and then carefully evaluating the contractor's proposals. Once the contract is signed, any changes are likely to prove costly, and the client has little further opportunity to comment on how the requirements are to be met.

It may be possible to modify a traditional 'work and materials' agreement where it is desirable to make the contractor responsible for design of part of the works. For example a Contractor's Designed Portion Supplement is available for use with the Standard Form of Building Contract (JCT98). Work such as piling, roof trusses etc.

could well be made the subject of such an agreement, and even though the contractor might wish to sub-contract such work, he would nevertheless accept responsibility for its design. The Supplement is a device which should be used as intended: that is, to modify the standard lump sum agreement. It should not be thought of as a way of introducing design obligations for a substantial part of the works, thereby effectively changing the original contract into one of a design and build type.

For management procurement

Basically such contracts concentrate primarily on the management expertise of the contractor and are particularly suitable for fairly large projects with complex requirements. There are many variants, ranging from the procurement of a building designed by the client's professional team on the one hand, to in addition playing a major design role, to also accepting a wider facilities management obligation to fit out and maintain the operation of a building for some specific period, and even perhaps playing a substantial role in the immediate funding of the project. Clearly in the majority of such cases, developers will have to look to forms of contract specially drafted to suit their requirements.

There are some standard forms available for use with the main management procurement options, as follows.

Management contracts

The client appoints an independent professional team, and also a management contractor. The contractor's involvement at pre-construction stages will be as an adviser to the team, and during construction he will be responsible for executing the works using direct works contractors.

With this type of contract it is possible to make an early start on site and achieve early completion. Because of its flexibility, it allows the client to develop the design during construction, because drawings and matters of detail can be adjusted and finalised as the work proceeds.

For a management contract to be successful there must be trust and good teamwork on the part of the client, the professionals and the management contractor. The latter should preferably be appointed no later than the outline design stage. He can advise on the design programme, tender action, delivery of materials and goods, and construction programmes.

The management contractor will normally make a written submission which includes a proposed management fee, and will be appointed after interviews with the client and the professional team. The fee will include for the total management service, expressed as a percentage of the total project cost, and for a service to cover pre-construction stages should the project not proceed to site.

The management contractor undertakes the work on the basis of a contract cost plan prepared by the quantity surveyor, project drawings, and a project specification. The client accepts most of the risk because there is no certainty about costs or programme. Competitive tenders for the works packages follow later, and they will usually, though not always, be lump sum contracts based on bills of quantities.

There is a Standard Form of Management Contract issued by the JCT, together with the Standard Works Contracts and an Agreement for use between the client and each works contractor.

Construction management

Again, the construction manager is appointed after a careful selection process and is paid a management fee.

One basic difference from a management contract is that the trades contracts, although arranged and administered by the construction manager, are direct between the client and the trade contractors. Although in a sense this gives the client a greater measure of control, it also means that he accepts virtually all of the risk. The construction manager is a coordinator, and usually cannot guarantee that the project will be finished to time or cost.

The client directs the project and the client is also likely to carry the greatest burden of the speculative risk. Obviously in-house expertise is essential.

Construction management seems to have increased in popularity over management contracts, but much work by this procurement method has been undertaken by experienced commercial clients using bespoke agreements. There are now also a few standard forms available for construction management.

Design, manage, construct and in some instances maintain

This variant is a more recent development of the management approach, offering a total integrated service from a single source. It can be led by the contractor, or by some other professional. The latter is only likely when the design aspects of the project are a high priority.

There are no standard forms available as yet, and developers or contracting organisations mostly use specially drafted forms. Architects involved in this kind of work should engage a specialist lawyer to check forms, and advise on the terms. Maintenance is normally outside the scope of building contracts, and is the subject of facilities management or other special agreements.

Contracts where the architect administers a series of separate trade contracts where there is no main or general contractor, may also fall into this category. As well as being the designer, the architect assumes the managing and coordinating roles for the project. This arrangement sometimes occurs with fairly small projects and is a quite traditional way of working. Trades contracts will ordinarily be direct between the client and the firms concerned.

Choosing the type of contract

In trying to decide which type of contract to adopt, an architect will have to ask him or herself (or others) a number of questions:

Q. *What is the nature or category of the work?*
Comment: is it predominantly building or engineering work? Is it a completely new building or an addition to an existing one? Is it refurbishment work, maybe of a specialist nature such as restoration of a historic building? Is it concerned with new uses for an old building, involving major alterations? Is it small jobbing work, perhaps on a 'one-off' basis, or one of a series of jobs which form part of a term maintenance programme?

Q. *Who is to be responsible for design?*
Comment: is responsibility to rest entirely with the architect as lead designer and the professional design team? Is the contractor to be directly involved in any important aspect of design? Is there to be significant design input from specialist sub-contractors or suppliers?

Q. *Will the nature of the work allow full and accurate tender documents to be prepared?*
Comment: it might be desirable to carry out exploratory work before going to tender, but this might not be possible in the event. With work to existing buildings, problems sometimes only become apparent after site operations have started.

Q. *Is there enough time to prepare full information at tender stage?*
Comment: does the urgent necessity for an early start on site mean that it will have to be made on the basis of notional or approximate information, to be replaced later with firm instructions?

Q. *What documents are needed for a particular type of contract?*
Comment: is this a project which can best be shown mainly in a drawn form? Are bills or schedules needed? Should the specification be a composite document which includes some descriptive sections, some scheduled and itemised, and some with quantities? What is then to be the status of the respective sections or documents in the event of conflict or discrepancies?

Q. *What is the method of selecting the contractor?*
Comment: is it to be by a direct negotiated approach, or by competition? Will competitive tendering be straightforward single stage, or is two stage tendering the only practical answer? This might affect the information to be prepared and will affect the time needed.

4 Which type of contract?

Q. Is the client able to state his requirements precisely before work starts?
Comment: is it desirable or necessary to allow for adjustments or design changes during the course of the work? What flexibility is needed and when are the critical times for final decisions?

Q. Does the client need to know a precise contract sum before work starts?
Comment: is construction cost of secondary importance to early completion (because, say, large sums of money are tied into property deals or the rental of existing premises, or because of dependence on economic returns)?

Q. Who is to be mainly responsible for coordinating the work on site?
Comment: is responsibility to rest primarily with the contractor, or does the client wish to reserve the right to nominate specialists, to arrange direct purchasing deals, or to have work carried out by others whilst the contractor is still in possession of the site?

Which Contract Form? 5

Decisions on the preferred procurement route will obviously have led to conclusions on the contract types which might be most suitable. Choice of the actual forms of contract to be used should depend on those circumstances surrounding the particular project, which need to be taken account of. The words and procedures described in the contract form need to adequately cover situations and foreseeable events which might arise during the carrying out of the work. The choice of form needs therefore to be based on an analysis of the intended work, and not be the result of prejudice, doctrinaire allegiance, or just lazily clinging to the familiar.

Some contract arrangements, particularly on major developments which might involve novel procurement or be projects of considerable complexity, could well require the use of specially drafted forms.

For most building operations in which the majority of architects are likely to be involved, one or more of the standard forms of contract should prove satisfactory. However, the tendency in recent years has been to introduce option clauses and supplements to make these forms more adaptable for use in a wider range of situations. The need to respond to recent legislation and changing practices in the construction industry, has also brought about a considerable increase in the number of forms which can be categorised as being standard. As a result there is now a considerable choice on offer, and the decision about which form, or even which combination of options is likely to be the most appropriate, becomes increasingly difficult to make.

Availability

Q. ***Is there a published standard form which will satisfy the requirements, or will a specially drafted document be needed?***

If extensive amendment is needed, or if a specially drafted document seems advisable, then the matter should be referred to the client for appropriate legal advice. An exception might be a small job in very straightforward circumstances which could be adequately covered by an exchange of letters. Even so, such a task should never be undertaken lightly, and a standard form which uses widely understood terminology might well be preferable. The industry has recently addressed the need for agreements which can be used by consumers on domestic work where no consultants are involved during construction, and the JCT home owners contracts, for example, are true consumer contracts. In all cases care is needed to ensure that contract provisions satisfy the requirements of relevant legislation.

5 Which contract form?

Q. What standard forms are published, and which is likely to be the most appropriate?

Chapters 6 to 13 below set out and compare, under a series of headings, the main features of the forms most commonly in use today.

Q. What if the contractor wants to use his own form of contract?

The client should be advised to pass any such document to his legal advisers for an opinion. Even simply worded and apparently clear documents can be heavily weighted in favour of the contractor's own interests, and the risks can rest largely with the client.

Q. What if the client wants alterations made to a standard form?

This might in effect make the contract non-standard and it could then be construed as contra proferentem. The basic rule is never to make amendments which might have unintended effects on other clauses and so upset the balance of the whole document. The Latham Report recommended that all parties in the construction process should be encouraged to use standard forms without amendment. Where there are exceptional reasons for such amendments, the architect should make sure that the client is aware of and accepts the risks. Amendments should be dealt with by his or her legal advisers. Again, there is the practical point that ad hoc alterations often lead to inflated tender figures as the contractor prices the risks.

Sector

This can have a significant influence, and is sometimes a matter over which there is little choice. This can be as a result of the need to comply with the standing orders of an authority, or to follow some official guidance on procedural matters. Several standard forms of contract were published in alternative versions, for use by either the private or the public sector as applicable. The present trend is for forms in only one version which can be used regardless of sector, and the distinction between what is permissible for private sector and public sector use is now becoming more fluid.

Government contracts

Until fairly recently construction work for government departments was invariably carried out under one of the GC/Works contracts, then the responsibility of the Department of the Environment. Today the situation is less clear cut.

For a start, the European Union now identifies three categories of authorities: central government departments (which seem to include NHS Trusts); sub-central government bodies (which include local authorities, police authorities, universities etc.); private bodies subsidised by government (e.g. Arts Council or English Heritage grant funded projects). Secondly, there are apparently no absolute rules about whether these authorities should use the GC/Works contracts in preference to other forms and the question is left fairly open-ended. The Latham Report clearly wished to see their use reduced in favour of the New Engineering Contract, but there is a considerable body of opinion in favour of continued use of the GC/Works forms. It would appear that most departments have been given discretion over whether to use JCT, NEC or GC/Works forms.

The GC/Works range of forms has been extended to suit most procurement methods, and all the forms have been extensively revised to take account of Latham Report recommendations and recent legislation. They are still published by The Stationery Office, and the forms can be used for work in both the public and the private sectors.

Local authority requirements

The public sector (or, more accurately, local authorities) has for many years had its own version of the JCT Standard Form. However, now that the person administering the contract is referred to as 'the Architect/the Contract Administrator', and some provisions peculiar to local government have become redundant, the need for private and public versions has been reduced.

For example, the JCT Intermediate Form (IFC98) shows that a one version publication is quite practicable for use regardless of sector.

In work for the public sector, there may also be need to take into account the published requirements of the particular department or client body. Some have firm views on the choice of forms, what amendments they insist are made and procedures which are to be followed concerning contract administration.

Familiarity

It is important, when considering an appropriate form of contract, to be thoroughly familiar with its provisions. Familiarity should never of course be the justification for choosing it. It takes time and effort to keep abreast of changes to published forms and any consequential effects on contract administration procedures.

Vigilance is also required to make sure that the editions of the contract forms and any addenda or supplements are the current issue and compatible. Checks are also advisable to make sure that the contract administration forms are current and compatible with the form of building contract being used. Pads of forms can quickly become out of date, and it is unreasonable to expect office staff to check these things when documents are issued. It is for the contract administrator to check such matters.

Personal preference

There is almost inevitably some subjective element in choosing a form of contract. An architect might feel a personal allegiance to one or other of the professional bodies. For example, ACA members might understandably incline towards an ACA form, while ASI members will feel a brand loyalty for their published forms. A majority of architects, mindful of the RIBA representation on the JCT and of its leading role historically, might instinctively turn to the JCT's range of standard forms.

Of course, equally it might be the client body which has strong preferences not only about the procurement method but also about the choice of form. Developers and members of organisations such as the British Property Federation might stipulate using forms which contain certain conditions. A project manager might advise the use of a form 'based upon' a well respected standard form (usually fairly loosely!), or a powerful funding body might require terms in a collateral agreement which necessitate considerable changes to those in the main contract.

Notwithstanding, the advantages of using a standard agreed form cannot be overstated. Modern forms drafted by experts and used without ad hoc alterations should result in fairer conditions and fairer tendering. Whatever the circumstances surrounding the choice of form, it should be the result of a thoughtful and logical analysis of the situation – an exercise in professional judgement.

Effect on contract for professional services

As discussed above, the choice of building contract form might also have implications for the contract for professional services. The Latham Report recognised this by its recommendation for a complete family of standard documents, including a total matrix of interlocking consultants' agreements and contracts. Publishers of some building contract forms also provide compatible sub-contracts, and ancillary documents including consultant agreements. However, the majority of consultants at present seem to prefer appointing documents published by their respective professional bodies.

The form of building contract chosen might have significance which should not be overlooked when calculating the fee for professional services relating to contract administration. For example, JCT98 demands fairly sophisticated procedures likely to be far more time-consuming than those needed under, say, IFC98. The type of contract, and the actual contract form which it is proposed to use, needs to be taken into account when compiling a fee bid – always assuming that the contract choice is known in sufficient time.

Even with apparently straightforward traditional lump sum contracts, there could be special factors which need to be borne in mind. For example, it might be expedient to arrange for a preliminary or enabling contract to cover investigation work, demolition work, advanced site works etc. ahead of the main contract, and this should not be overlooked when assessing fees; or again, it might be decided to have work carried out under a number of parallel trades contracts with the architect assuming a management and coordinating role, and charging an additional fee for this additional service.

The services provided and the fee appropriate will always depend on the nature of the work. For example, work on the preservation or restoration of historic buildings might warrant a considerable departure from the Outline Plan of Work stages, with a great deal of investigation and reporting needed in the earlier stages followed by unusually frequent and detailed inspections both before and during construction.

Another matter which often brings close links between the building contract and contract for professional services is design liability. The architect should only accept design liability to an extent agreed and intended and which is acceptable for professional indemnity insurance purposes. Care is needed not to acquire inadvertently any liability that is more onerous, for example, fitness for purpose as the result of incautiously entering into some collateral warranty with a tenant or purchaser, or when acting for a contractor involved in a package deal with strict liability.

Traditional procurement 6

Standard lump sum forms

The Joint Contracts Tribunal Ltd
Major Project Form 2003

The Joint Contracts Tribunal Ltd
Standard Form of Building Contract 1998 Edition

The Stationery Office
GC/Works/1 With Quantities (1998)

The Institution of Civil Engineers
NEC Document
Engineering and Construction Contract Second Edition

The JCT Major Project Form is placed first in the list because it is a new and significant approach. It seems likely that usually, though not necessarily, it will be for a lump sum contract. It may also be thought of as a form which is versatile enough for use with either traditional or design and build procurement.

It should be noted that of the other forms listed above, the last two are also suitable for use with procurement other than traditional lump sum. They are listed under this heading for convenience, and in recognition of their seminal position as major forms of contracts

JCT MPF03

The Joint Contracts Tribunal Ltd

Major Project Form 2003 Edition

Background

It appears that for some time the JCT Standard Form of Building Contract (JCT98) has no longer entirely satisfied the needs of some employers undertaking major commercial developments. The separation of design from construction is less of a clear cut reality than it traditionally was, and the employer often wishes the contractor to take responsibility for some detailed design, but at the same time requiring flexibility and allowing the employer to maintain a measure of control. The extended time scale of major projects has frequently necessitated proceeding in phases, in parallel working for both design and construction, in payments related to progress and performance, and with incentives for savings in time and cost. In the past this has often resulted in employers making extensive modifications to the conditions in otherwise standard forms of contract, or using bespoke forms.

Now thanks to a bold initiative by the JCT, a new standard form is available which takes account of the current needs of this market. It is quite a significant departure from anything previously published by JCT, and in format, style and language, it breaks new ground. In some respects it might seem to be something of a hybrid, although direct comparisons with JCT98 or WCD98 are not altogether helpful. However, it should certainly be regarded as a likely alternative to those well tried documents, if used in the particular circumstances for which it is intended. In other circumstances the two long established forms might still offer an appropriate option, and presumably they will continue in print alongside the Major Project Form.

As a document it is very open and flexible but demands such attributes of its users. The relative brevity, straightforward language, and clear procedures make it immediately attractive. However it is likely to be an appropriate choice only for parties who are already experienced in operations of a major scale, and who fully appreciate the nature of the conditions and procedures. They may be relatively brief and uncomplicated, but they depend on a high level of understanding and involvement by the parties.

Nature

MPF03 is much shorter than JCT98. This relative brevity is due in part to the use of enabling clauses which avoid the inclusion of that which might be irrelevant for the particular project, and allow the parties to include their own detailed project-specific requirements on matters such as insurance.

It is logically structured, but interestingly there are no conventional Articles or Recitals, and the document goes immediately into the 39 Contract Conditions which are set out under eight section headings. The Conditions conclude with an unusually full set of definitions (including the meaning of Practical Completion) and are followed by the Attestation. Understandably, this is a contract to be executed as a deed.

The Appendix requires project-specific information to be entered, and the choice of option clauses to be indicated. Tables require entries relating to completion by sections, and insurances.

The document also includes a Third Party Rights Schedule in respect of third party rights from the contractor in favour of the funder, and third party rights from the contractor in favour of a purchaser or tenant; and the Pricing Document. This last document includes the rules which govern the manner of payment to the Contractor, and pricing information. Details relating to an advance payment bond, Progress Payment Schedule, and Contract Sum Analysis must be attached where relevant.

The form has a number of interesting key features including:

- the Contractor is responsible for any additional design work required beyond that shown or described in the Employer's Requirements;
- the design submission procedure ensures that the Employer is invited to review and comment on all the Contractor's design documents;
- the Employer may name specialists and sub-contractors, and the Contractor becomes responsible for their performance;
- the Employer may novate those consultants previously appointed and the Contractor then becomes responsible for their services performed pursuant to the Requirements;
- the Employer gives the Contractor access to the site, rather than possession;
- insurance arrangements are for the parties to agree upon;
- interim payments to the Contractor will usually be made monthly, but may be to any terms the parties agree. There is no provision for retention, but bonds or security measures similar to retention could be created if required;
- there are provisions for acceleration, bonus for early completion, and sharing of benefits of savings or value improvements;
- the Employer must appoint a sole representative who has authority to act in all matters under the contract. Other consultants appointed by the Employer can expect cooperation from the Contractor, but will have no authority under the contract to act for the Employer.

For the purpose of comparisons, in this book the synopsis is in the same format as that employed for notes on the other forms of contract, although this will not directly reflect the structure used in the actual contract form. Clause numbers cited are those to be found in MPF03.

JCT MPF03

Use

The form is published in one version only, and in theory is therefore available for use in both the private and public sectors. In practice it is likely to be used mainly with large commercial developments, but might also have relevance for PFI projects.

It is for a lump sum contract, and the Contract Sum is to be stated in the Appendix. The Employer's Requirements could include bills of quantities, but there is no specific reference to these in the document. The Contractor's Proposals must be accompanied by a Contract Sum Analysis and Pricing Information. The Requirements and the Proposals are key documents on which the agreement is founded.

There is no provision for an independent and impartial Contract Administrator, and administrative functions required under the contract are to be taken direct by the Employer or the Contractor. The Employer's Representative may exercise all the powers and functions of the Employer.

Synopsis

1 Intentions

- The Contractor is obliged to execute and complete the Project, that is for construction works and the completion of the design (1·1).
- The Contractor warrants that he or she has the competence and resources to act as Planning Supervisor, Principal Contractor, and designer as required under the CDM Regulations (1·3).
- The Employer is responsible for the Requirements, and the Contractor is not responsible for the contents of these, or for the adequacy of design contained in these (5·1).
- If a discrepancy within the Requirements is found, the Contractor must notify the Employer which provision it intends to follow. If the Employer wishes the Contractor to proceed otherwise, that instruction is treated as a Change (4·2).
- If a discrepancy within the Proposals is found, the Employer will instruct the Contractor which provision should be adopted and that instruction is not treated as a Change (4·3).
- Where the Contractor takes over design, he or she warrants that he or she will perform to the standard of reasonable skill and care appropriate to a competent professional, and does not warrant fitness for purpose (5·3).
- Should it be thought necessary to stipulate a fitness for purpose obligation, then the JCT Major Project Form Guidance Note suggests suitable wording, and also gives a reminder of some of the practical problems likely to arise with this kind of provision (footnote to 5·3).
- Materials and goods are to be of the kinds and standards described in the contract,

JCT MPF03

and where not described are to be reasonably fit for the intended purpose (5·4).

- Where described goods and materials are not procurable the Contractor must propose alternatives of an equivalent or better standard. Only if the alternative is of a lesser kind or standard, and acceptable to the Employer, will it be treated as a Change (5·4).
- Workmanship is to be of the standards described or otherwise is to be executed in a good and workmanlike manner (5·5).
- The Contractor is responsible for the preparation of further design documents, and these are to be submitted to the Employer for review in the quantities and format all as identified in the Appendix, and as shown on the design programme contained in the Requirements or Proposals (6·2).
- The Employer is to respond by returning design documents to the Contractor marked either 'A Action', 'B Action' or 'C Action' (6·3).
- The Contractor is to execute work marked 'A Action' and can expect to be paid accordingly (6·6·1).
- The Contractor is to execute work marked 'B Action' provided that the Employer's comments are incorporated, and a further copy of the document is immediately submitted to the Employer. The Contractor can expect to be paid accordingly (6·6·2).
- The Contractor must not execute work marked 'C Action', but must resubmit a document which takes account of the Employer's comments (6·6·3).
- Regardless of comments made by the Employer and subsequently incorporated into the design, the Contractor is still responsible for ensuring that any design document he or she prepares is in accordance with all the requirements of the contract (6·10).

2 Time

- The Contractor is to be given access to the site (note not possession) on the date stated in the Appendix and must proceed regularly and diligently to achieve Practical Completion on or before completion date (9·2).
- The contract allows for completion by sections, and completion dates for each section (together with rates of damages and bonus) are to be entered in the Appendix Table.
- With the consent of the Contractor, the Employer may take over any part of the Project prior to Practical Completion, and must then issue a statement identifying the part, the date, and the value of the part (11·2).
- The Contractor must use reasonable endeavours to prevent or reduce delay to progress or to completion (9·3).
- Whenever the Contractor becomes aware that progress of the Project is being or is likely to be delayed due to any cause, he or she must forthwith notify the Employer

of the cause of the delay and likely effect upon completion of the Project (12·2).

- Unless there in anything to the contrary in the contract, the Contractor will be entitled to an extension of time in respect of eight listed events. Of these four are for 'neutral causes', and the other four relate to any Change, interference by other persons on site, valid suspension of performance for non-payment, and acts of prevention by the Employer. Interestingly, adverse weather is not listed (12·1).
- Where the Contractor considers that delay is caused by one of these events, he or she must provide supporting evidence, and revise this as necessary (12·3).
- Within 42 days of receipt of such notification, the Employer must notify the Contractor of such adjustment to the completion date as he or she then considers fair and reasonable, or give reasons why the completion date should not be adjusted (12·4).
- The Employer may review his or her decisions at any time in the light of further documentation (12·5).
- Within 42 days after Practical Completion of the Project, the Contractor may provide further documentation to support any further adjustment to the completion date, and within 42 days of receiving that information the Employer must review decisions made previously, and either confirm or adjust the completion date (12·6).
- The Contractor must notify the Employer when in his or her opinion Practical Completion has occurred and, if he or she agrees, the Employer will issue a certificate of Practical Completion (9·4).
- If the Contractor fails to achieve Practical Completion by the completion date, he or she becomes liable for liquidated damages at the rate stated in the Appendix (10·1).
- No adjustment to the completion date to bring an earlier completion date is possible except by agreement (12·8).
- However, the Employer can investigate the possibility of acceleration by inviting proposals from the Contractor. The Contractor must either make proposals or explain why it is impracticable to achieve an earlier date (13·1).
- If the date of Practical Completion is earlier than the completion date, then the Employer is liable to pay the Contractor a bonus at the rate entered in the Appendix (14·1).

3 Control

- The Contractor may not assign either the benefit or burden of the contract without the consent of the Employer, and this would apply both to construction work and to design work (29·1).
- However, the Employer may assign the benefit of the contract without consent (29·2) and furthermore the Contractor consents to the Employer assigning both the benefit

JCT MPF03

and burden of the contract to the Funder (named in the Appendix) at any time (29·3).

- Rights of third parties in general are excluded, but a Third Party Rights Schedule forms part of the contract, in respect of the Funder, and a purchaser or tenant (30·1).
- The contract has no provision for a conventional Contract Administrator and the terms are to be administered direct by the Employer and Contractor respectively. However, the Employer is required to appoint an Employer's Representative who will exercise the powers and functions of the Employer under the contract (15·1).
- The Employer may appoint other advisers who, although assured of cooperation from the Contractor, have no authority under the contract to act for the Employer (15·2).
- Instructions from the Employer must be in writing and must be empowered under the contract. The Contractor must comply with such instructions (2·1).
- Where the Contractor fails to comply, then subject to seven days' notice in writing, the Employer may engage others to give effect to the instruction and the Contractor is liable for the extra cost (2·3).
- Where an instruction gives rise to a Change, this may entitle the Contractor to additional payment and an extension of time. However, not all instructions will be treated as giving rise to a Change, and where the contract states this, the Contractor will not receive additional payment or time. This will not relieve the Contractor of any obligations under the contract (2·2).
- The Employer may instruct the Contractor to open up and test work and materials. If the findings show all to be in accordance with the contract, this will constitute a Change. If the work, materials, or goods are not in accordance with the contract, it will not be a Change (16·1).
- Where work, materials or goods are not in accordance with the contract then the Employer may instruct their removal, or may allow them to remain but with a price reduction. The Contractor will not be entitled to loss and/or expense or extension of time. The Employer may also instruct additional work necessary as a consequence, and may also instruct further opening up and testing relating to similar work materials or goods elsewhere (16·2). No such instructions shall be treated as a Change (16·3).
- During the 12 months following Practical Completion (termed the Rectification Period), the Employer may instruct the Contractor to remedy any defect.
- Where the Contractor does not remedy defects as instructed, then the Employer may engage others to give effect to the instruction (17·1).
- After expiry of the Rectification Period, and with all defects rectified, the Employer shall issue a certificate to that effect (17·2).
- Consultants may be pre-appointed by the Employer with the intention that the

appointment will be novated to the Contractor, and full details of their original appointment and the proposed Model Form of Novation should be included in the Requirements. This also requires an appropriate Appendix entry (18·1).

- The Model Form of Novation must be executed immediately upon entering into MPF03 (18·2).
- The Employer may also require the Contractor to appoint named specialists (that is, subcontractors or consultants) by including a name, or list of names from which the Contractor may chose, in the Requirements. Such appointments may be in respect of design or carrying out of works (18·3).
- The Contractor becomes solely responsible under the contract for services provided by any pre-appointed consultant, and work undertaken by named specialists (18·4).
- Payment to the Contractor by the Employer depends upon the novation of pre-appointed consultants and the appointment of specialists being carried out in the manner required under the relevant contract clauses (18·5).
- Contracts with pre-appointed consultants or specialists may not be amended or terminated by the Contractor without the prior written consent of the Employer (18·6 and 18·7).
- Either before or immediately following termination of such contracts, the Contractor must notify the Employer of the proposed replacement, and the Employer has seven days in which to raise reasonable objection (18·8).
- The Contractor remains fully liable under the contract for replacement consultants or specialists, and entirely responsible for any delay and additional cost incurred (18·10).

4 Money

- The Contract Sum, VAT exclusive, is entered in the Appendix (25·1).
- The Pricing Document identified in the Appendix is part of the contract, and should contain the rules which will determine the method of payment, the Contract Sum Analysis, and pricing information such as rates, preliminaries, and overheads which can be used in the valuation of Changes.
- Changes are alterations in the Requirements or Proposals which affect either the substance or manner of what the Contractor is to provide under the contract. As the Contractor is responsible for all further design information, then excepting discrepancies and statutory requirements, Changes will be mainly due to variations required by the Employer.
- Each party is to notify immediately the other if it is considered that an instruction gives rise to a Change, or any event occurs which should be treated as giving rise to a Change (20·1).

JCT MPF03

- Valuation of a Change and any adjustment of the completion date may be by agreement, or by the Employer on the basis of a quotation provided by the Contractor, or, if no quotation is received or agreed, on a fair valuation basis. The valuation is to be inclusive of any loss and/or expense (20·3 to 20·6).
- No later than 42 days after Practical Completion of the Project the Contractor may provide particulars of any further valuation in respect of any Change, and within 42 days of receipt of the particulars the Employer must review relevant previous valuations (20·9).
- Other factors which may result in adjustments to the Contract Sum include amendments to the Requirements and Proposals as suggested by the Contractor which will be cost savings and value improvements resulting in a financial benefit to the Employer (19), and a bonus payable for early completion (14). The former will be the proportion of any benefit calculated as stated in the Appendix, and the latter will be at the daily rate entered in the table in the Appendix.
- The Contractor has only limited rights to reimbursement of loss and/or expense outside the inclusive valuation of Changes. If the Contractor wishes to claim, it must give timely notification and provide an assessment of the loss and/or expense with information updated as necessary. Within 14 days of receipt of information, the Employer must ascertain the amount and notify the Contractor. Payment will be included in the next payment advice. Any claim for further ascertainment must be made by the Contractor within 42 days after Practical Completion of the Project, and the Employer must review these particulars and notify the Contractor of any additional payment appropriate within a further 42 days (4·20).
- Relevant deletions in the Appendix will indicate whether the Contractor is to receive payments under Rule A (interim valuation), Rule B (stage payment), Rule C (progress payments), or Rule D (some other method). These Rules are described in the Pricing Document. If no Rule is selected then A will apply.
- The Contractor is to make a detailed application for payment not later than seven days before payment is due, and the Employer is to issue a payment advice on the day of each month as entered in the Appendix (22·1).
- After Practical Completion the Employer will issue further payment advice at monthly intervals, always provided that the amount due is not less than the figure entered in the Appendix (22·2).
- Each interim advice is to state the amount due to the Contractor (determined in accordance with the Pricing Document), the value of any Changes, and the amount of any reductions. It should be noted that there is no provision for payment of unfixed materials or cost fluctuations, although these could be incorporated if desired (22·5).
- There is no express provision for retention.
- The Employer may withhold payment due to the Contractor provided that effective

notice is given not later than seven days before the final date for payment, and which states the grounds for withholding and the amounts attributable to each ground (23).

- If payment is not made in accordance with the contract, interest of 5 per cent over base rate will become payable on outstanding amounts (24·1).
- Final payment becomes due when rectification of defects has been certified, or when the Employer considers that reasonable time has been allowed for rectification to have taken place. The Employer must issue a Final Payment Certificate, and this is final and binding on the parties in relation to amounts due, subject to any dispute being referred to adjudication or litigation within 28 days (22·7).

5 Statutory obligations

- Statutory requirements are defined as Acts of Parliament, local authority and statutory undertaker's regulations and byelaws, and any directive of the European Community having the force of law (39·2).
- The Contractor is to comply with all statutory requirements and warrants that the design of the Project (except for that contained in the Requirements) complies with statutory requirements (5·2).
- The Contractor is to make any applications and give any notices required by statute and will pass copies of relevant documents to the Employer (3·1).
- Unless the Requirements state that specific fees and charges are the responsibility of the Employer, the Contractor will pay all fees or charges in connection with statutory requirements (3·2).
- Changes in statutory obligations that arise after the Base Date and were not previously announced must be taken into account and will be treated as giving rise to a Change (4·5).
- The Contractor is appointed as both Planning Supervisor and Principal Contractor for the purposes of the CDM Regulations (1·2). The Appendix also allows for the name of a Planning Supervisor previously appointed by the Employer to be entered, and if this person is to be retained then the contract would need amending (1·2).
- The Contractor also warrants that it has the competence and resources to fulfil the roles of Planning Supervisor, Principal Contractor, and designer for the purposes of the CDM Regulations (1·3).

6 Insurance

- The Contractor indemnifies the Employer in respect of personal injury or death and damage to property other than the Project, always assuming that these arise in the course of carrying out the Project and are not due to some act or neglect for which the Employer has responsibility (26·1).

JCT MPF03

- The Employer indemnifies the Contractor against expense, liability, loss, claim or proceedings arising under statute or common law in respect of personal injury or death and damage to property other than the Project, always assuming that these arise in the course of carrying out the Project and are due to some act or neglect for which the Employer has responsibility (26·2).
- Most major projects are likely to need bespoke insurance arrangements and this contract requires the relevant documents to be attached with details identified in the Appendix (27·1).
- Either party may be required to provide and maintain cover, and the other party may request documentary evidence (27·2). Failure to provide satisfactory evidence within seven days will allow the other party to take out insurance and recover the costs involved (27·3).
- Where compliance with the Joint Fire Code is a contract provision and the insurer requires remedial measures, the Contractor must implement these and this is not treated as giving rise to a Change (27·1).
- Where cover against terrorism is required but ceases to be available, the party responsible for that insurance must notify the other party (27·7). The risk then rests with the Employer (27·8).
- Where professional indemnity insurance is required, a relevant deletion is required in the Appendix, and the limit of indemnity is to be entered. The Contractor may be required to take out and maintain cover until 12 years from the date of Practical Completion of the Project, always assuming that cover remains available at commercially reasonable rates (28·2).

7 Termination

- The Employer may, by issuing a further notice, terminate the employment of the Contractor if, after having given the Contractor 14 days' notice of a material breach, the Contractor has failed to remedy the breach (32·1 and 32·2). The Contractor's employment may also be terminated in the event that the Contractor becomes insolvent (32·3).
- Material breach by the Contractor is defined, and includes failure to proceed regularly and diligently, failure to comply with an instruction, suspension of the Project, breach of the CDM Regulations, breach of provisions relating to named specialists or pre-appointed consultants (39).
- Upon termination, the Contractor must provide the Employer with all design documents, and must not remove any materials, plant or equipment from site without permission. The Employer may then make appropriate arrangements to complete the Project (32·4). Only when the Project has been completed or, if no other arrangements for completion have been made, within six months of termination, must the Employer

issue a payment advice (32·5 and 32·7).

- The Contractor may, by issuing a further notice, terminate his or her employment if, after having given the Employer 14 days notice of a material breach, the Employer has failed to remedy the breach (33·1 and 33·2). The Contractor may also terminate its employment in the event that the Employer becomes insolvent (33·3).
- Material breach by the Employer is defined, and is confined to failure to issue a payment advice in accordance with the contract (39).
- Upon termination, the Contractor must remove all his or her materials, plant or equipment from site without delay, and prepare an account of amounts due (33·4).
- Either party may terminate the Contractor's employment if the Project is substantially suspended for the period stated in the Appendix due to causes which include force majeure, specified peril, civil commotion, and terrorism (34·1).
- Upon termination the Contractor must provide the Employer with all design documents, remove all materials, plant or equipment from site without delay, and prepare an account of amounts due (34·4).

8 Miscellaneous

- Definitions and meanings are fully tabled. Note in particular Design Documents, Model Form, Practical Completion, Requirements and Proposals (39·2).
- The Contractor is given access (not exclusive possession) to the site or parts of the site, which leaves the Employer free to have work undertaken by others at the same time as the Project (9·1).
- Ground conditions or man-made obstructions encountered by the Contractor will only give rise to a Change where they could not have been foreseen (8·2).
- Copyright in all design documents prepared by the Contractor remains vested in the Contractor and the Employer is given an irrevocable licence to use them for the purposes of the Project. Where the Contractor does not own the copyright in any design document he or she shall procure a licence from the copyright holder (7·1 and 7·2).
- The stated period in which acts are to be done will commence immediately after the specified date. Christmas Day, Good Friday and bank holidays are excepted (39·1).
- All communications between the parties relating to the contract are to be in writing, or may be made electronically by the procedures specified in the Appendix (38·1). Any notice under the Third Party Rights Schedule or relating to termination must be given by actual delivery, registered post or recorded delivery (38·2).

JCT MPF03

9 Disputes

- Disputes or differences between the parties in relation to the Project (note not simply the customary 'arising under this contract') may be submitted to mediation if the parties agree, or referred to adjudication in accordance with the provisions of the relevant Scheme for Construction Contracts (36·1 and 37·1).
- Although the contract states that the objective of mediation should be to reach a binding agreement, the only final resolution of disputes would appear to be by legal proceedings. There is no provision for arbitration.
- The law of the contract is to be the law of England (35·2).

JCT MPF03

This contract?

If considering using MPF03 remember that:

It is intended for use where both the Employer and the Contractor, together with their respective teams of specialists and sub-contractors, are experienced in substantial commercial projects. The employer is required to appoint a Representative who will exercise all the powers and functions of the Employer under the contract. Other advisers may be appointed but they will have no authority under the contract to act on behalf of the Employer.

The Requirements and the Proposals are at the heart of this contract, and it is important that both are full and explicit. There is no prescribed format but it should be remembered that this contract is very reliant on the Requirements stating clearly what is to be delivered, and the manner of delivery. Specific points which may be considered for inclusion are helpfully listed in the JCT Guidance Note for this contract.

The contract provides for design by the Employer as shown and described in the Requirements, with further design by the Contractor. The Requirements may stipulate that the Contractor engages consultants pre-appointed by the Employer under a novation agreement. The JCT Guidance Note lists some of the matters which need to be covered in the novation agreement, which should become part of the Requirements. Normally design liability is that of reasonable care and skill, but the Requirements could include a fitness for purpose obligation although there might be practical difficulties. (This might be an important matter in the case of PFI projects.)

Completing the Appendix requires entries on matters such as the Contract Sum, Requirements, Proposals, Pricing Document, names of Planning Supervisor, Funder, adjudicator, entries relating to the application of option clauses such as those for ground conditions, liquidated damages, bonus rates, pre-appointed consultants and named specialists, cost savings, payments, insurances, professional indemnity, and communications.

This is a form which is simpler and shorter than either JCT98 or WCD98, and which can be tailored to the needs of the project. However, it should be approached with respect, because although the openness and apparent brevity of the Conditions is admirable, professional and sometimes legal advice might be advisable to produce a reasonably balanced set of documents. The Contractor assumes more risks and responsibilities than under other JCT forms of contract, but provided that the risks can be fully identified and priced for at tender stage, this should not present a problem for experienced operators. What effect this admirable document will have on the use of the more traditional JCT forms remains to be seen!

JCT MPF03

Related matters

Documents

JCT Major Project Form 2003
Subcontract

References

JCT Major Project Form: Guidance Note

Commentaries

Sarah Lupton
Guide to MPF03
RIBA Enterprises (2003)

JCT 98

The Joint Contracts Tribunal Ltd

Standard Form of Building Contract 1998 Edition

Background

The original agreed Standard Form was published in 1909 and was known as the RIBA Form. In many respects JCT98 is a direct descendant through a series of editions published in 1931, 1939, 1963 and 1980, although now publication is the responsibility of The Joint Contracts Tribunal Limited (JCT).

JCT80 was drafted to overcome various deficiencies which had appeared over time in the 1963 Edition, and in response to recommendations in the Banwell Report – in particular those relating to the treatment of sub-contractors. At the time it was felt that JCT80 had struck a fair balance between the interests of the contracting parties and others involved. Initially concern was expressed by some who saw the form as being much longer than its predecessor and more demanding to administer, yet it quickly gained acceptance as the form to be used for major building projects in the United Kingdom. Supplements were soon available to cater for contractor's design, sectional completion, together with documentation for nominating sub-contractors and suppliers.

JCT80 was subject to 18 Amendments, and Amendment 18 was an attempt to meet many of the recommendations in the Latham Report of 1994, and to ensure compliance with Part II of the Housing Grants, Construction and Regeneration Act 1996 in respect of adjudication and payment provisions. The 1998 Edition of the Standard Form was basically a consolidated version of JCT80 but since its publication has received four Amendments.

Nature

JCT98 runs to over 100 pages. The Articles of Agreement include Recitals, Articles and Attestation. The numbering of Recitals and Articles may vary depending on the particular edition of JCT98, and if incorporating any separate Amendments great care needs to be taken to achieve consistency.

The Conditions are set out in five parts, and some parts will not be applicable in all cases. Part 1 includes the general conditions which will apply to all contracts. Part 2 covers Nominated Sub-contractors and Nominated Suppliers which may or may not be applicable. Part 3 gives the alternatives for dealing with Fluctuations, although the actual detailed clauses are published separately. Part 4 is Settlement of Disputes, and includes Adjudication, Arbitration and Legal Proceedings. Part 5 is Performance

JCT 98

Specified Work. Also incorporated is a Code of Practice relating to clause 8·4·4 concerning work not in compliance with the contract. In addition to an Appendix there is also Annex 1 on Bonds, Annex 2 on EDI provisions, and a Supplemental VAT Agreement.

Obviously such a contract is not an 'easy read' and the number of options need to be considered with great care. For example, the Appendix requires very careful consideration of whether certain clauses come into the applies/does not apply category. The form lacks a logical structure, because it has evolved over many decades in a rather ad hoc manner. Many of the clauses include not only legal conditions but also detailed procedures and rules. Some of these, especially those concerning sub-contractors or payment, might appear to be arduous but they are intended to secure sound practice and should be followed meticulously.

Metric numbering of the clauses is unremarkable, but the numbering of some sub-clauses is taken to many decimal points. When referring to such clauses, e.g. in correspondence or instructions, care is needed to ensure that the right number of points appear in the right places!

Use

The form is published in six versions making it suitable for use in the private sector or by local authorities, and for use with quantities, without quantities, or with approximate quantities. A Supplement is also available to make the local authorities' version suitable for use by government departments.

The With Quantities version should only be used where the Employer, through his professional consultants, has provided at the time of tender a full set of drawings and bills of quantities to SMM7. An Information Release Schedule is part of the documentation in an attempt to identify responsibility for any further information which might be necessary to amplify the contract during the carrying out of the Works.

The Without Quantities version also requires preparation of a full set of drawings to be accompanied either by a Specification or Schedules of Work. In order to give valuation of variations and fluctuations a substantive basis the Contractor is also required to submit a Schedule of Rates or a Contract Sum Analysis; this should be provided, and the measure of detail required of the Contractor is often stipulated at tender stage by the Employer.

JCT98 requires the appointment of a person to give effect to the various contract terms. The Employer will usually appoint an Architect or a Contract Administrator to this role, and such a person will be regarded as independent and impartial. However, the local authority version of JCT98 also provides for the Employer to appoint an 'Employer's Representative' (1·7) to administer the contract, instead of an Architect/Contract Administrator. A footnote emphasises the fundamental difference in roles.

JCT 98

Synopsis

1 Intentions

- The Contractor is obliged to carry out and complete the Works in accordance with the contract documents (2·1). The Contractor's responsibility is clearly reinforced (1·4) (1·5 in Private version) and remains wholly his.
- The quality and quantity of work included in the Contract Sum is that set out in the bills (14·1).
- The contract documents are to be read as a whole (1·2) and the printed Articles, Conditions and Appendix prevail (2·2·1).
- In the case of discrepancies in or divergences between documents (including nominated sub-contractors' Numbered Documents) corrective instructions must be given (2·3).
- Contract bills, except where specifically stated otherwise, must be prepared in accordance with the Standard Method of Measurement, 7th Edition (2·2·2).
- The Contractor must be provided with two copies of the information referred to in the Information Release Schedule by the stated times (Second Recital and 5·4·1). The Contractor must also be provided with such further drawings or details which are reasonably necessary (5·4·2). All documents issued must be returned after completion, if requested (5·6), and use of them by the Contractor is limited. Confidentiality of rates and prices must be respected (5·7).
- The Contractor is to supply copies of his master programme as soon as possible (5·3). (Many architects require a preliminary draft with the tenders.)
- The CDM Regulations oblige the Employer to nominate a Planning Supervisor and where relevant a Principal Contractor. This is also a contractual duty (Article 6) and any need to appoint a replacement is also covered.

2 Time

- Dates for possession and completion should be entered in the Appendix. The Contractor must proceed regularly and diligently and complete on or before completion date (23·1·1). Early completion is an option for the Contractor and, if achieved, issue of the Practical Completion certificate cannot be delayed. Conversely, the Employer is not obliged to assist the Contractor in attempts to complete early.
- An option clause for deferment of possession not exceeding six weeks may apply, subject to an Appendix entry (23·1·2).
- Notice of delay must be given in writing by the Contractor, together with supporting information including his estimate of the likely effect on completion (25·2). The Architect is required to consider a new completion date, and to notify the Contractor

JCT 98

of his or her decision within 12 weeks (25·3). Nineteen 'relevant events' are listed which are grounds for an extension of time (25·4) the most recent of which relates to any impediment, prevention or default by the Employer. The interim decision is subject to review by the Architect no later than 12 weeks following Practical Completion. Whilst it is possible to reduce extensions already awarded, the original contract period cannot be reduced and there is no provision for accelerating progress. The procedures for dealing with delay and extensions of time are detailed and need to be followed with care.

- Failure by the Contractor to complete within the contract period is certified simply as a fact by the Architect (24·1) and liquidated damages may be deducted or otherwise recovered by the Employer (24·2). The Employer is obliged to give notice in writing beforehand.
- Practical Completion is certified by the Architect (17·1).
- After this, the Contractor is obliged to rectify defects (17·2) unless the Employer decides otherwise and takes an appropriate deduction instead.
- There is provision for partial possession (18·1), and where the Employer wants to use part of the uncompleted works for storage etc., this is possible subject to proper insurance arrangements (23·3).

3 Control

- The bar to assignment without written consent refers to 'the contract' (19·1·1). Sub-contracting any part of the work requires the Architect's written consent (19·2), and any sub-contractor, other than a nominated sub-contractor is a domestic sub-contractor.
- Subsequent options under JCT98 are a little complicated. There is a straightforward authorised domestic sub-let; the more restricted choice by the Contractor from a list of names; and the non-domestic nominated sub-contractor.
- The Contractor may sub-let to persons named in a list included in or annexed to the contract bills. There must be not less than three firms 'able and willing' to carry out the specified sub-contract work at the required time (19·3). Architects who use this might consider whether they require the Contractor to indicate choice at the time of submitting his tender, in order to discourage so-called 'Dutch' auctioning.
- JCT98 also has provision for nominating sub-contractors and suppliers (35 and 36). The clauses are of necessity very detailed and the procedures involved need careful study and timely application. Forms of tender, warranties and sub-contract conditions are expressly referred to (35·4) and are published separately as NSC/T, NSC/A, NSC/N, NSC/W and NSC/C, respectively. There are precise obligations concerning renomination, non-completion to time, Practical Completion and direct payment for nominated sub-contractors.

JCT 98

- There is an option clause which, if an Appendix entry states that it is to apply, allows the Employer to transfer a right of action against the Contractor to persons with a subsequent interest in the completed works (19·1·2).
- Architect's instructions must be in writing, although this can mean written confirmation of oral instructions (4·3). The contract clearly defines what instructions are empowered, and these may include variations (13) expenditure of provisional sums (13·3), and postponement of work (23·2).
- The Contractor is required to have a competent person-in-charge on the site full time (10), and to permit the presence of the Employer's Clerk of Works. The Clerk of Works is solely an inspector, although he can issue directions which require confirmation by the Architect (12).
- Where work or materials do not comply with the contract, the Architect can order their removal from the site (8·4). Where, after consultation, it is agreed that non-conforming work should remain, then the Employer is entitled to an appropriate deduction. The Architect is empowered to order tests and inspection (8·3) and the likelihood of any non-compliance in similar work elsewhere is covered (8·4·4 and Code of Practice).
- The contract requires all work to be carried out in a proper and workmanlike manner, and in accordance with the Health and Safety Plan (8·1·3). In the event of failure to comply, and although this might under other circumstances be interference with the Contractor's working methods, the Architect is empowered to issue instructions (8·5).
- The contract does allow for work under the direct control of the Employer to be carried out during the time that the Contractor is in possession (29).

4 Money

- This is an area where a number of amendments were introduced into JCT80, largely the result of the Latham Report and subsequent legislation. These resulted in conditions and procedures significantly more stringent than previous. In principle, the emphasis is on placing initiative with the Contractor to price instructions and variations; more rigorous procedures concerning payments to aid cash flow; time limits for payments beyond which interest becomes due, and valid grounds for suspending work in the event of non-payment. You ignore these at your peril.
- The Contract Sum is VAT exclusive (15·2) and may only be adjusted as provided for in the Conditions (14·2).
- The Conditions provide for the Contractor to at least recover increases in taxes, including landfill tax, levies or contributions promulgated after the date of tender (38). Increases in the cost of labour and materials may be recovered as net increases (39) or in accordance with Formula Rules (40), if such an option is included.

JCT 98

- Where provisional sums have been included, instructions must be given to the Contractor (13·3). In accordance with SMM7, provisional sums will be either for defined or for undefined work. The valuation of work carried out where an approximate quantity is included, or where provisional sums are included, or where a variation has been ordered, is to be on the basis of a Contractor's Price Statement (Alternative A) i.e. his application of the valuation rules; or Alternative B i.e. the quantity surveyor's application of the valuation rules.
- In addition, the Contractor may be invited to submit a '13A Quotation' for work which is the subject of an Architect's instruction. Such a quotation, if accepted, would bind the Contractor to the direct cost of work, the time implications and any loss and/or expense which might apply. The work would not be carried out on this basis unless a confirmed acceptance was issued. In the event that the quotation is not accepted, an instruction may still be issued to proceed with the work, but it will then be subject to the valuation rules and procedures (13·5).
- The Contractor must make written application for reimbursement of loss and/or expense. The grounds for any valid application are set out (26·2) and include only matters over which the Contractor has no control and which occur because of action or failure by the Employer. The procedures should be followed precisely, and the Contractor's proper written notice and supporting information is a requirement (26·1).
- Private versions of JCT98 include advance payment to the Contractor where this is stated in the Appendix to apply (30·1·1·6) and this might be subject to an advance payment bond. The sum is to be reimbursed to the Employer in agreed amounts and at agreed times.
- Interim payments to the Contractor can be at pre-arranged stages (30·2) but are more likely to be at intervals as entered in the Appendix (usually monthly). Interim valuations will be made by the quantity surveyor (30·1·2·1). but may arise from application by the Contractor and his own assessment of the gross valuation (30·1·2·2).
- Interim certificates must show the amount due, and must also state the basis of calculation. Within five days of issue the Employer must notify the Contractor in writing of the amount of payment he intends to make. Any intentions to withhold or deduct must be clearly stated in a written notice which also sets out the grounds for such action. This must be given to the Contractor no less than five days before the date for final payment. The final date for payment is 14 days from the date of issue of the Interim Certificate, and if no valid notices are given by the Employer then payment must be made in full. Failure to pay by the final date for payment will attract interest of 5 per cent over current base rate, and can also give the Contractor a right to suspend work (30·1·4).
- Where the traditional operating of retention is to apply, in preference to any form of bonding, then this will apply to all Interim Certificates (30·2). Half the retention

amount will be released at Practical Completion. Certification becomes more complicated when there are nominated sub-contractors, because the amounts included for each sub-contractor must be identified separately.

- There are precise procedures to be followed in the preparation of the final account after Practical Completion (30·6·1). Issue of the Final Certificate is within two months of sending the final account to the Contractor (always assuming that work to be carried out at the end of the Defects Liability Period has been satisfactorily dealt with (30·8)). Similar rules apply to the Final Certificate with regard to notices, as those outlined above for Interim Certificates, except that the final date for payment is 28 days from the date of the certificate.

5 Statutory obligations

- It is the Contractor's duty to comply with all statutory obligations and give all required notices (6·1). He is entitled to recover fees and charges not otherwise provided for (6·2).
- The Contractor is to notify the Architect if he finds any conflict between statutory requirements and the contract documents (6·1·2). The Architect must issue an instruction and the Contractor is thereafter not liable to the Employer under the contract for any non-compliance with statutory requirements resulting from the instruction (6·1·5).
- The Contractor is empowered to carry out limited work for emergency compliance and this will be treated as a variation to be valued accordingly (6·1·4).
- Where the CDM Regulations apply, the Contractor is contractually obliged to comply with all the relevant CDM duties, particularly in respect of the Health and Safety Plan and in providing information for the Health and Safety File (6A).

6 Insurance

- What a particular contract includes will depend to a large extent on Appendix entries (for example, whether option clauses are to apply, the minimum amount of cover required, etc.). The Architect may be obliged to issue instructions, call for documentary evidence, and pass to the Employer for checking.
- The Contractor indemnifies the Employer in respect of personal injury or death, and injury or damage to property other than the actual works (20). This is to be backed by insurance (21·1) and the minimum amount of cover required is an Appendix entry.
- If instructed, the Contractor is to take out joint names insurance for the Employer against the risk of claims arising due to legal nuisance. There is a list of exceptions, and damage must not be attributable to any negligence by the Contractor. An Appendix entry will indicate whether cover may be required (21·2), and the amount of cover to be provided.

JCT 98

- Insurance of 'the Works' is for all risks where new buildings are concerned and should be for full reinstatement value. It can be taken out either by the Contractor (22A) or by the Employer (22B). Normally it is better to leave any risk with the Contractor under 22A, since restoration under 22B is treated as variation work and will be valued accordingly.
- Insurance of existing structures and the contents is a matter for the Employer (22C) and is limited to specified perils. New work in existing buildings, although still a matter for the Employer, requires all risks cover.
- Insurance may be required against the Employer's loss of liquidated damages due to an extension of time following damage to the Works (22D). This requires an Appendix entry, but experience to date suggests that cover is not easy to obtain and is likely to prove expensive.
- An Appendix entry will show whether the Joint Code of Practice on the Protection from Fire of Construction Sites is to apply (22FC) and if so, both Employer and Contractor must comply with it. In the event of non-compliance, the insurers can specify remedial measures which must be undertaken. In the event that terrorism cover is withdrawn and is no longer available, the situation and options open to the Employer are dealt with in clauses 22A·5, 22B·4, or 22C·5 as applicable

7 Termination

- The Employer is allowed to determine the employment of the Contractor by reason of specified defaults (27·2). A warning notice may be issued by the Architect, but the notice of determination is a matter for the Employer. In the case of insolvency of the Contractor, depending on the circumstances, determination might be automatic subject to possible reinstatement, or the Employer might enter into an agreement with the Contractor (a '27·5·2·1 agreement') for continuation or novation (27·5).
- Depending on the circumstances the Employer can have the right to make interim arrangements for certain work to be carried out during the holding period of a 27·5·2·1 Agreement, or to have the Works completed by another Contractor (27·6), or to decide not to have the works carried out and completed at all after determination of the Contractor's employment (27·7).
- The Contractor is allowed to determine his own employment for specified defaults by the Employer (28·2). Again, the procedures must be followed meticulously. In the event of insolvency of the Employer, the Contractor may elect to determine his own employment.
- Either party can determine the employment of the Contractor for listed neutral causes (28A).
- The respective rights and duties of the parties concerning payment, removal and completion are set out (27·6, 28·4 or 28A·2).

JCT 98

8 Miscellaneous

- A list of definitions relevant to JCT98 is included (1·3).
- There is a contracting out of third party rights under the Contracts (Rights of Third Parties) Act 1999 (1·12).
- Access for the Architect is covered (11) but this might be subject to reasonable restrictions as far as workshops are concerned.
- The Architect has power to order the exclusion of persons from the Works (8·6).
- Where progress is disturbed because of the discovery of antiquities, the Contractor is obliged to inform the Architect and to take all necessary action to preserve the status quo and avoid disturbance. The Architect must issue instructions, and the Contractor is entitled to ascertained loss and/or expense (34).

9 Disputes

- Part II of the Housing Grants' Construction and Regeneration Act 1996 gives either party a statutory right to refer any difference or dispute arising out of the contract to adjudication. Article 5 of JCT98 provides for this.
- Procedures for referral to adjudication, the appointment of an adjudicator, the powers and conduct of an adjudicator are set out (41A).
- The adjudicator's decision is binding on the parties at least until the dispute is finally determined at arbitration or by legal proceedings.
- Article 7A establishes arbitration as an agreed method of resolving disputes, provided the Appendix shows that clause 41B applies.
- The appointment of the arbitrator is subject to an Appendix entry, and his powers are defined (41B).
- The parties agree that either may apply to the courts on a question of law (41B·4).
- Arbitration is to be conducted in accordance with the JCT 1998 Edition of the Construction Industry Model Arbitration Rules (41B·6), and the provisions of the Arbitration Act 1996 shall apply (41B·5).
- Where the Appendix states that clause 41B does not apply, then any dispute or difference is to be determined by legal proceedings, and the court is given extensive powers to open up Architect's certificates etc. (7B).

JCT 98

This contract?

If considering using JCT98 remember that:

It is intended for substantial lump sum contracts and is available for use with or without quantities and in different versions for private and public sector use. Work needs to be fully designed and documented in detail at tender stage, and is for completion within a stated period. The Employer is required to appoint a Contract Administrator and a quantity surveyor.

If used for work in Northern Ireland an Adaptation Schedule should be incorporated, while for work in Scotland the Scottish Building Contract version of the form should be used (see Chapter 7 below).

It can include partial possession, sectional completion, Contractor's designed portion and performance specified work. It allows for sub-contractors to be chosen by the Contractor from a list of not less than three names, and for nominated sub-contractors and suppliers. Use of dedicated documents is mandatory with nominated sub-contractors.

When completing the form, entries are required relating to decisions on matters including deferment of possession; bonds (whether in lieu of retention, advance payment or 'listed items'); insurance of the Works; Joint Fire Code; liquidated damages; advance payment; fluctuations; and EDI. Care is needed to ensure that the relevant Supplements are used, and option clauses selected.

If acting as Contract Administrator note that some of the procedural rules are detailed and likely to prove time-consuming, particularly where there are nominated sub-contractors and suppliers.

Amendments are issued by the JCT from time to time. The form is available on disk. The RIBA publishes contract administration forms for JCT98.

This form is a direct descendant of the 1909 first standard form and now tending to show its age. It is a sophisticated heavyweight and with Supplements can make for a bulky document. It is not logically structured, and the sometimes lengthy conditions are not always easy to grasp. However, the format, language and terminology have become familiar to many over time, and there is a considerable body of case law. It is probably still the most widely used form for major building work.

JCT 98

Related matters

Documents

Standard Form of Building Contract JCT98:
Private With Quantities
Private Without Quantities
Private With Approximate Quantities
Local Authorities With Quantities
Local Authorities Without Quantities
Local Authorities With Approximate Quantities
Amendment 1: 1999 (Construction Industry Scheme)
Amendment 2: 2000 (Sundry amendments)
Amendment 3: 2001 (Terrorism cover/Joint Fire Code/SMM)
Amendment 4: 2002 (Extension of time/loss and expense/advance payment)

Nominated sub-contractor forms:
NSC/T98: Part 1 (tender invitation, architect to sub-contractor)
NSC/T98: Part 2 (tender)
NSC/T98: Part 3 (particular conditions of tender)
NSC/A98 (agreement, contractor and sub-contractor)
NSC/N98 (nomination from architect to contractor)
NSC/W98 (warranty agreement, employer and sub-contractor)
NSC/C98 (conditions of sub-contract)
(all these documents have to be used if nominating sub-contractors)

Nominated supplier forms:
TNS/1Tender
TNS/2Warranty
JCT Domestic Contracts: (no requirement in contract that these particular forms be used)
Sub-contract Agreement DSC/A/SC 2002 Edition
Sub-contract Conditions DSC/C/SC 2002 Edition

Supplements:
Fluctuations: Private (includes with and without quantities)
Fluctuations: Local Authority (includes with and without quantities)
Sectional Completion Supplement (with quantities /approximate quantities/without quantities)
Contractor's Designed Portion Supplement (with quantities/without quantities)
Composite Contractor's Designed Portion and Sectional Completion (with quantities/without quantities)

References

Earlier JCT Practice Notes (relevant to JCT98 but written with JCT80 in mind):
Practice Note No 23: Contract Sum Analysis (1987)
Practice Note No 24: Insolvency of Main Contractor (1992)
Practice Note No 25: Performance Specified Work (1993)
Practice Note No 27: Application of the CDM Regulations (1995)
Practice Note No 28: Mediation (1995)
Series 2 JCT Practice Notes (yellow covers):
Practice Note 1: Construction Industry Scheme
Practice Note 2: Adjudication (includes text of agreements)
Practice Note 3: Insurance, Terrorism Cover
Practice Note 4: Partnering (includes text of non-binding charter)
Practice Note 5: Deciding on the Appropriate JCT Form of Contract
Practice Note 6: Main Contract Tendering (includes model forms)

Commentaries (relating to JCT98)

David Chappell
Parris's Standard Form of Building Contract
3rd edn, Blackwell Science (2002)

Sarah Lupton
Guide to JCT98
RIBA Publications (1999)

JCT in collaboration with Building Design Partnership
Guide to the Use of Performance Specifications
RIBA Publications (2001)

RIBA Publications
Architect's Guide to the Contract Administration Forms JCT98
RIBA Publications (2000)

GC/Works/1 (1998)

The Stationery Office

GC/Works/1 With Quantities (1998)

Background

GC/Works/1 first appeared in 1973 as a form intended for use almost exclusively by central government departments. GC/Works/1 (1998 Edition) is a direct successor and is published for use in major civil engineering or building projects, under traditional, design and build, or management procurement. It can also be used by non-central government agencies and private clients.

The form is part of a family of GC/Works contracts, and GC/Works/1 is available in six versions as follows:

GC/Works/1: With Quantities (1998)
GC/Works/1: Without Quantities (1998)
GC/Works/1: Single Stage Design and Build (1998)
GC/Works/1: Two Stage Design and Build (1999)
GC/Works/1: With Quantities Construction Management Trade Contract (1999)
GC/Works/1: Without Quantities Construction Management Trade Contract (1999)

GC/Works/1 contracts are for use on major building and engineering projects. The range of contracts under the GC/Works label is comprehensive and further includes:

GC/Works/2: for minor building and engineering works
GC/Works/3: for mechanical and electrical engineering works
GC/Works/4: for mechanical and electrical small works
GC/Works/5: for the appointment of consultants
GC/Works/6: for a daywork term contract
GC/Works/7: for measured term contracts
GC/Works/8: for a specialist term contract for equipment maintenance
GC/Works/9: for operation, repair and maintenance of plant, equipment and installations
GC/Works10: for facilities management

GC/Works/1 (1998) With Quantities is a particularly complete publication and can be adapted to suit a wide range of applications. It is similar in structure to its immediate predecessor and even uses the same numbers for most of the Conditions. There are fundamental differences, however, in the text which are not always immediately apparent. The earlier published form was intended almost exclusively for use by government departments and reflected the methods and procedures of contract administration then used by them. The form was not intended to be even-handed in all matters, and the Project Manager was afforded absolute authority with

GC/Works/1 (1998)

many of his or her decisions being 'final and conclusive'. The current form is claimed to be adaptable enough for use by non-central government employers (for example local authorities, educational institutions, housing associations, NHS Trusts, etc.) and by private sector employers. To facilitate this an attempt has been made to produce a form which strikes a fair balance between the interests of the Employer on the one hand and those of the Contractor on the other. A 'fair dealings' clause has been introduced, and there is recognition of the fact that the Project Manager's decisions are now open to adjudication.

The form takes account of the recommendations in the Latham Report, and complies with the conditions of the Housing Grants, Construction and Regeneration Act 1996 (Part II).

Nature

The document runs to over 80 pages. There is an introduction and contents list, followed by the Conditions which are set out in a clear graphic style using straightforward language and well established terminology. Also included are a Schedule of Time Limits (useful summary for contract administrators); a detailed alphabetical index; the customary Abstract of Particulars but with an Addendum for entries about information yet to be supplied; and the tender forms. The Contract Agreement is in two versions so that the contract can be under Scots law, or under the law of England, Wales and Northern Ireland.

Use

The form is intended for use in major building or civil engineering works.

The Conditions include:

- design by the Contractor and sub-contractors;
- professional indemnity insurance for design;
- incentive bonus for early completion;
- finance charges;
- mobilisation payments;
- payments to the Contractor on the basis of stages, milestones or valuations;
- performance bonds;
- parent company guarantee;
- collateral warranties.

The factual details relating to a particular contract and the incorporation of option provisions will be determined on how the Abstract of Particulars is completed. The Abstract is detailed and among other things requires the names of the Project Manager and Planning Supervisor (who may be the Project Manager) to be entered. There is also space for the adjudicator and the arbitrator to be named in this document. It is recommended in the notes that the same adjudicator and arbitrator are named in

GC/Works/1 (1998)

all the Employer's related contractual documents, whether with the Contractor, consultants or others. This could be problematic if the disputes are not related.

A Contract Agreement is to be executed in duplicate and the date entered. Government contracts are not normally executed under seal, but there is space for attestation if required.

Synopsis

1 Intentions

- There is a fair dealing clause which requires both parties to act in good faith and in a cooperative and open relationship (1A).
- The Contractor is to execute the Works with diligence, in accordance with the Programme, with all reasonable skill and care, and in a workmanlike manner. If any part of the Works does not conform with the contract and is rejected by the Project Manager, then it must be replaced by the Contractor at his own expense (31[6]). The terms 'the Works' and 'Things' are defined (1[1]).
- The Contractor may be required to undertake responsibility for design work in respect of such work carried out by himself, or by a sub-contractor or supplier. His liability can be either that of the professional duty to exercise reasonable skill and care (10 – Alternative A) or to give a fitness for purpose warranty (10 – Alternative B).
- The Contractor is deemed to have satisfied himself about the conditions under which he will work (7[1]). No additional payment is allowed except for unforeseeable ground conditions (7[5]).
- The quality of 'Things' for incorporation is to conform to the requirements of the Specification, bills and drawings (31[2]). The Contractor must be prepared to satisfy the Project Manager in respect of the execution of the Works, and that he is using the skill and care expected of an experienced and competent contractor.
- The 'contract' means the Contract Agreement, Conditions, Abstract of Particulars, Specification, drawings, bills of quantities, Programme, tender, and the Employer's written acceptance (1[1]).
- If discrepancies occur between Specification and drawings, or between drawings, the Contractor is to draw the Project Manager's notice to any discovered (2[3]). The Conditions prevail where documents conflict with them (2[1]).
- Bills of quantities are to be prepared in accordance with the method of measurement identified, except where stated otherwise (3[1]). Any errors or omissions in the bills are to be rectified by the Employer (3[3]).
- The Contractor is to receive a copy of drawings issued 'during the progress of the Works' in a form which the Project Manager considers suitable for reproduction (2[5]).

GC/Works/1 (1998)

2 Time

- The contract period will be stated in the Abstract of Particulars (34[1]). The Employer will notify the Contractor when he may take possession of the site or parts of the site. All notices under the contract are to be in writing (1[3]).
- The Contractor is required to proceed with diligence and in accordance with the Programme, or as the Project Manager instructs (34[1]). The whole of the Works or any relevant section must be completed in accordance with the contract, and to the satisfaction of the Project Manager by the date for completion (34[1]).
- The date for completion is set out in or ascertained from the Abstract of Particulars (1[1]). This envisages Practical Completion (although not called such) and includes clearing of rubbish and all Things not incorporated (34[2]).
- The Contractor is required to submit a Programme prior to acceptance of the tender, for it to be agreed by the Employer (1[1]). The Programme is to be for the whole period for completion, and must show sequence and other specified information (33[1]).
- Regular progress meetings are to be held (35[2]), and the Contractor is obliged to notify the Project Manager of delays and anticipated delays (35[3]). If notice is given, or if the Project Manager is already aware of likely delay, he or she shall consider whether or not to award an extension (36[1]). The causes for which an extension may be awarded are listed (36[2]), and the Project Manager is to indicate whether the award is interim or final. The Project Manager is to keep interim decisions under review until a final decision is possible (36[3]). It is interesting that weather is not recognised as a cause of delay.
- The Project Manager is required to issue a written statement of progress within seven days after each progress meeting (35[4]).
- Acceleration of completion is possible upon direction by the Employer, subject to acceptance of Contractor's priced proposals (38). The Contractor may also choose to submit priced proposals and Programme amendments for the Employer to consider.
- The Project Manager shall issue a certificate when the Works, or any section, are completed in accordance with the contract (39).
- Failure to complete the Works or a section (which includes clearance) by the relevant Date of Completion makes the Contractor liable to the Employer for liquidated damages (55). There is no reference to a certificate of non-completion.
- The Maintenance Period will be stated in the Abstract of Particulars (21[1]). The contract accepts that there might be more than one Maintenance Period. The Contractor is obliged to make good defects at his own cost and to the satisfaction of the Employer. Any arguments about liability and reimbursement must wait until after defects have been rectified.

GC/Works/1 (1998)

- There is provision for completion of the Works by section if specified in the Abstract of Particulars (1[1]).
- There is provision for the Employer to take early possession of any part of the Works, and this also relates to completion by sections (37).

3 Control

- The Contractor is not allowed to assign or transfer the contract or any interest without written consent of the Employer (61). Sub-letting is also barred without prior consent of the Employer or the Project Manager (62[1]).
- In any sub-contract, the Contractor is required to ensure certain terms (62[2]). The Contractor is responsible for seeing that sub-contractors comply with all obligations imposed upon them. The Main Contractor must see that sub-contract works are completed (62[4]).
- Sub-contractors or suppliers may be nominated on the basis of a Prime Cost sum (63[1]). In the event of determination of a nominated sub-contract the Employer may renominate a replacement or direct the Contractor to complete the work (63[7]). There are no stated procedures and no requirement to use a particular form of contract. The Contractor is given right of reasonable objection (63[6]).
- Instructions from the Project Manager must be in writing (40[3]) except for a few (listed) which can be oral and confirmed later. The contract sets out what instructions are empowered (40[2]).
- Instructions can be given by the Project Manager or delegated to his or her representative (4), and the Contractor must comply forthwith. Instructions requiring a variation are termed 'VIs'. The Project Manager may require the Contractor to submit a quotation of the full cost of complying with a VI within 21 days of the instruction (40).
- In the event of failure to comply with the Project Manager's instruction, the Employer may have the work done by others at the Contractor's expense (53). This right extends to rectifying defects (21[3]).
- The Project Manager is to provide the Contractor with information necessary for setting out the Works, and the Contractor is solely responsible for the correctness of the setting out (9).
- The Contractor shall employ a competent agent (5) who is to supervise the Works, be in attendance at site during all working hours, and supply the Project Manager with returns (15). A Clerk of Works or Resident Engineer may be appointed, and the Project Manager or quantity surveyor may appoint representatives to exercise their respective powers – which must be listed (4).
- The Project Manager may inspect, examine or have tests carried out (31[4]). Independent experts may be brought in to test fitness or suitability of Things, and if

GC/Works/1 (1998)

their findings disclose non-compliance with the contract, then the Contractor must bear the cost of rectification and any necessary further tests (31[5]).

- The Employer has power to execute other works (which may or may not be in connection with the Works) during the execution of the contract (65).

4 Money

- The Contract Sum is defined (1[1]) and may be adjusted as provided for in the contract.
- Provisional sums require instructions in writing from the Project Manager before work under these items begins (64). Valuation is as provided for in the contract (42).
- Valuation of Project Manager's instructions (40) may be by acceptance of a lump sum quotation. If there is no agreement, the quantity surveyor will value on the basis set out (41) or (42[5]) in the case of a variation instruction, or (43) in the case of other instructions. Prolongation or disruption costs may be included as part of the valuation.
- The right to prolongation and disruption expenses generally is limited (46[1]). The matters are set down in the Conditions. Interest and finance charges are expressly excluded (46[6]). Recovery of expenses depends on written application from the Contractor made to the Project Manager in time (46[3]). The application must meet the requirements set down.
- Finance charges may be payable to the Contractor only for limited reasons, and for stated periods (47[1] and [3]). The rate is to be stated in the Abstract of Particulars (47[2]).
- Progress payments, termed 'advances on account' (48), are based on either Stage Payments, Milestone Payments or Valuations (50[2]). Payments will include for work executed to the satisfaction of the Project Manager, and the Contractor is entitled to 95 per cent of the relevant sum, plus 100 per cent of certain other sums and certain adjustments (48[2]). Stage Payment Chart and Milestone Payment Chart are defined (1[1]).
- After completion of the Works, the Contractor is entitled to be paid the amount estimated by the Employer as the Final Sum, less half the retention. The quantity surveyor shall send a copy of the final account to the Contractor within six months of certified completion (49[2]). The Contractor must notify agreement or otherwise within three months. The other half of any retention is released when the Final Certificate is issued at the end of the longest Maintenance Period, and when the Contractor has complied with making good defects.

5 Statutory obligations

- The Contractor is to give all statutory notices required, obtain any consents necessary, and pay fees and charges arising. The Employer will reimburse fees or charges properly

GC/Works/1 (1998)

incurred. An obligation on the Contractor acting as Principal Contractor to comply with the CDM Regulations is expressly stated (11).

- The Contractor is also required to conform to any occupier's rules or regulations (if stated in the Abstract of Particulars) relevant to the site or the premises within which he is working (22).
- The Contractor is required to comply with all statutory requirements which govern the storage and use of all Things brought on to the site (13[2]).

6 Insurance

- The Contractor is required to maintain for the duration of the Contract and the longest Maintenance Period: employers' liability insurance; insurance against loss or damage to the Works and Things for full reinstatement value; insurance against personal injury or damage to property (8). The evidence can be required within 21 days from acceptance of tender (Alternative A). The Contractor may be required to maintain insurance in the joint names of the Employer, the Contractor and all sub-contractors in accordance with details attached to the Abstract of Particulars (Alternative B).
- Where so stated in the Abstract of Particulars, and in connection with a Contractor's design obligation, he may be required to take out and maintain professional indemnity insurance cover (8A).
- The Contractor is responsible for loss or damage to the Works and even extending to any Things not for incorporation in the Works (19[6]). This is in respect of any loss or damage, but where this arises because of 'accepted risks', defined in (1[1]), unforeseeable ground conditions or unforeseeable circumstances beyond the control of the Contractor, the Contractor will be reimbursed by the Employer. There is an absolute obligation on the Contractor to reinstate, replace or make good to the satisfaction of the Employer (19).
- The Contractor is to take precautions needed to take care of the site and the Works against loss or damage from fire, and any other cause, and shall take all reasonable steps for security and protection of the site and Works including lighting and watching (13[1]).

7 Termination

- The Employer may determine the contract at any time by giving notice to the Contractor. This is a discretionary power which can be 'at will', and not conditional upon some default by the Contractor (56).
- In addition, the Employer has the right to determine the contract for specific defaults by the Contractor including insolvency (56[6]).
- The Contractor may determine the contract on various stated grounds, and matters following determination by the Contractor are covered in Condition 58.

GC/Works/1 (1998)

- Matters following determination by the Employer (e.g. payment, completion, removal, transfer of sub-contracts) are dealt with in Condition 57. The quantity surveyor shall ascertain and the Project Manager certify the cost to the Employer of completion of the Works.

8 Miscellaneous

- A list of definitions is included (1[1]).
- There appears to be no express condition to contract out from the Contracts (Rights of Third Parties) Act 1999.
- The Contractor is to give the Project Manager reasonable notice before covering up work (17) and shall not lay foundations until the Project Manager has examined the excavations (16).
- The Project Manager may order the replacement of the Contractor's site staff, including the agent (6[1]). The Project Manager has power to control the admittance of certain persons to the site (26).
- Security measures such as a requirement for passes (27), the taking of photographs (28), and awareness by all employees of the Official Secrets Acts (29) may be an obligation, where stated in the Abstract of Particulars.
- The Contractor has express obligations relating to the protection of the Works (13), the prevention of nuisance (14) and the removal of rubbish (34[2]).
- There is provision for the discovery of antiquities (32[3]).
- There is an extremely wide provision for recovery of sums where money is owed by the Contractor or to the Contractor, under this or any other contract with the Employer (51).

9 Disputes

- There is provision for adjudication for the resolution of any dispute arising during the course of the Works (59[1]). There are precise requirements for the notice of referral and the procedures. A decision may normally be expected within 28 days of the notice. The adjudicator's decision is binding until the dispute is finally determined by legal proceedings or by arbitration.
- Arbitration is included in addition to adjudication as a means of dealing with disputes (60[1]), and the arbitrator is given wide powers under the contract.

6 Traditional procurement: standard lump sum forms

GC/Works/1 (1998)

This contract?

If considering using GC/Works/1 remember that:

This was the major form for lump contracts originally drafted for use by central government, but it has been substantially revised in the 1998 Edition to allow for a much wider application, including use by private sector employers. However, it is still intended primarily for use by government departments or Crown agencies. There is a version for use with quantities, another for use without quantities.

It can be used for work in England and Wales, and for work in Northern Ireland or Scotland. In the latter cases Condition 60 deals with the differences in arbitration resulting from the contract being subject to Northern Ireland law and Scots law. There are relevant references to statute law which applies to these countries, and a different Contract Agreement for use under Scots law is included. Fuller helpful information is given in the Commentary under 'Legal Background'.

It can include completion in stages, design by the contractor with professional indemnity insurance, security measures, early possession, acceleration and cost savings, bonuses, nomination of sub-contractors or suppliers, mobilisation payment, alternatives for advances on account, and bonds in lieu of retention. The wording is clear, with a good graphic layout. The procedures are logical, and there are many interesting features in the provisions not to be found elsewhere.

When completing the contract details, the Abstract of Particulars is a vital document with which to tailor the particular contract to the intended works.

If acting as Project Manager, contract administration should be relatively straightforward. The obligations and responsibilities are clearly stated, the person concerned is given considerable authority, and the procedures are not arduous. However, it is necessary to keep a careful watch on the Schedule of Time Limits.

The 1994 Latham Report recommended that government departments then using GC/Works/1 should begin to change to the New Engineering Contract. The current 1998 GC/Works/1 family of forms may be seen as a robust response to that suggestion. It has emerged as a versatile and well structured document which embodies many of the Latham Report's points for 'an effective form of contract in modern conditions'. The Government Central Advice Unit also publishes some excellent Information Notes from time to time.

Related matters

Documents

GC/Works/1 With Quantities (1998) General Conditions

GC/Works/1 Without Quantities (1998) General Conditions

GC/Works/1 Model Forms and Commentary

GC/Works Sub-Contract

NEC

The Institution of Civil Engineers
New Engineering Contract Document

Engineering and Construction Contract Second Edition

Background

This form is placed under the heading of Traditional Procurement and Lump Sum Forms purely for reasons of convenience. It is actually part of a system which was a bold and major initiative in the drafting of construction contracts. This resulted in a form which is adaptable and claimed to be suitable for use in lump sum, design and build, or management procurement, and for both civil engineering and building works.

The Engineering and Construction Contract (ECC) is a Second Edition development of what was first called the New Engineering Contract (NEC). This was an entirely new 'clean sheet' approach to drafting construction contracts undertaken for the Institution of Civil Engineers. It was prepared by a panel of engineers and lawyers chaired by Dr Martin Barnes of Coopers and Lybrand, London.

The NEC was an attempt to offer:

1. Flexibility: by making one all purpose document suitable for traditional procurement, design and build, or management contracts. It was further claimed that it could be suitably adapted for most types of civil engineering and building work, from large-scale projects down to domestic-scale work. Although drafted as a head contract, it was originally thought that it could also be used as a sub-contract, thus providing the ultimate in back to back compatibility. The language of the form was such that it was thought suitable for use under UK law and also overseas.
2. Clarity: sufficient for the form to be exportable, understandable, and therefore likely to lead to fewer disputes. It was also drafted with plain language and relatively short clauses kept in mind. The precise meaning of some of the more unfamiliar terminology, however, has yet to be defined by case law.
3. Good management on the part of all parties: it was felt that there should be an end to adversarial posturing by bringing into the contract an obligation for frank and open discussion of problems as they arose, thereby minimising the risk of disputes escalating to the point where time and costly overruns became inevitable. It was further thought desirable to introduce incentives for good performance and early completion.

Sir Michael Latham bestowed high praise on the original New Engineering Contract

NEC

in his 1994 Report. After listing what he considered to be desirable features which should be present in all modern construction contracts, he stated that 'the approach of the New Engineering Contract is extremely attractive' and that it contains 'virtually all these assumptions of best practice'. He went on to advocate certain changes to the New Engineering Contract, some of which were as follows:

- that the name should be changed to New Construction Contract (hence the change to ECC);
- that there should be provision for a secure trust funding;
- that there should be prompt payment provision for sub-contractors;
- that there should be affirmation by the parties that all dealings were to be on a fair basis;
- that core clauses should be left unamended and only compatible sub-contracts used;
- that terms of appointment for consultants and adjudicators should interlock with the contract;
- that consideration should be given to a short and simpler minor works variant (now available).

The Latham Report also recommended that 'Government Departments should begin changing to the NEC' and that 'the use of NEC (amended) by private sector clients should be strongly promoted'. Since the Latham Report was published, with its 13 principles for a modern contract, the Second Edition of the form has appeared, and has been amended to meet many of the Latham Report's recommendations. Reports indicate that the NEC has been taken up widely, and that it is now being used on projects in various sectors of construction and engineering in various parts of the world.

The publication of the ECC also seems to have spurred a response from the JCT and the publishers of the GC/Works forms. They have produced amendments to their documents which take up many of the recommendations of the Latham Report, and are obviously keen to keep abreast of trends. The ICE continues to publish its well established more conventional contract forms as well as ECC documents.

Nature

The ECC system comprises 13 documents, which include a professional services contract and an adjudicator's contract, together with relevant guidance notes. The ECC is intended as a reference document, and contains Core clauses and Optional clauses.

The Core clauses are set out in nine sections, and apply in all contracts. The Main Option clauses constitute six sets of clauses A to F, and one set will be added to the Core clauses to adapt the document to the type of contract required (for example Priced Contract With Quantities or Activity Schedule; Target Contract; Cost Reimbursable Contract; Management Contract). The Main Option clauses A to F are also published, with the Core clauses, as separate documents.

NEC

There are also Secondary Option clauses which may be incorporated or not as required. Some can only work in conjunction with certain Main Option clauses. There are 15 headings (G to Z) which can be incorporated as required to allow for performance bonds; guarantee bonds; advance payment; sectional completion; limitation of Contractor's liability for design; fluctuations; retention; bonus for early completion; delay damages; low performance damages, etc. This gives great opportunity to tailor the conditions to the intended works.

The ECC is a novel piece of drafting in 'simple language' but non-traditional terminology could cause some problems over interpretation. The use of the present tense is also rather disconcerting and might exacerbate problems for lawyers construing the terms.

In an attempt to stimulate good management, emphasis is laid on the contractual importance of effective programming, sound management, and the need for early warnings by both the Project Manager and the Contractor in order that matters which have implications for progress and additional costs can be properly considered at the earliest possible time. In a sense it anticipated the present partnering trends, but a partnering option is now also available for use with this and other NEC documents.

Use

First, it is necessary to select the appropriate Main Option. Then the Secondary Options can be considered and incorporated as desired. It is necessary to determine the make-up of the contract content before completing the statements of Contract Data. The Core clauses are relatively brief, and the information carried in the Works Information and the Contract Data therefore becomes extremely important. Part One of the latter consists of information to be provided by the Employer, and Part Two is data provided by the Contractor. Because of the number of options available, and the fact that data will be related to the selected options, completion of the Works Information and the Contract Data needs to be precise and approached with great care.

Synopsis of Core clauses

1 Intentions

- The intentions of the parties as indicated in the contract documents should be evident from the Core clauses, the choice of Option clauses and the Contract Data provided by the Employer.
- The contractual spirit of mutual trust and cooperation is expressed in the 'General Section' (10·1) of the Core clauses.
- Duties of the Employer, Contractor, Project Manager and Supervisor are set out (11 and 14).
- There are clear rules relating to communications (e.g. instructions, certificates,

NEC

submissions, records, etc.) (13).

- Early warning is a significant requirement in the contract, although possibly a quite difficult concept in practice (16).
- Instructions are required from the Project Manager to resolve any ambiguities or inconsistencies (17).
- Health and safety matters (related particularly to the CDM Regulations) become contractual as well as statutory obligations on the Contractor (18).
- The Contractor is obliged to notify the Project Manager of work which he feels is either impossible to execute or illegal (19).
- The Contractor's main obligation is to provide the Works all in accordance with the Works Information (20), defined in clause 11·2. It is important therefore to make certain that this information is full, clear and accurate.
- The contract allows for design by the Contractor (21) and questions of design liability and copyright are dealt with (22).
- The Contractor is responsible for cooperation over providing information, and the sharing of working areas is a contractual obligation (25).
- Title to plant, equipment and materials is normally vested in the Contractor, but may pass to the Employer where the Supervisor marks goods and materials as being 'for the contract' (70). The subject of title is allocated a complete section of the Core clauses.

2 Time

- Starting and completion arrangements are straightforward (30). Possession of the site, access and use is subject to conditions (33).
- Programmes should be described in the Contract Data, or otherwise submitted to the Project Manager by the Contractor. The detail to be included is described (31). Programmes may be revised subject to conditions (32).
- The Project Manager is empowered to issue instructions that work stops, or that it is not started until instructed (34).
- Procedures for using or taking over part of the Works early by the Employer are subject to conditions. Possession of any part of the Works, or the whole site, is dependent upon certification by the Project Manager (35).
- The Project Manager can require the Contractor to submit a quotation for accelerating the Works in order to achieve completion ahead of the contract completion date (36).
- Where delay occurs due to certain intervening events, these may constitute Compensation Events (60). There are 18 listed, and for weather in particular, precise requirements can be included in the Contract Data. The Contractor is obliged to notify

the Project Manager of such events, and the Project Manager will decide whether compensation is due (61). Obviously the 'early warning' requirement will be taken into account. Compensation Events are afforded a complete Section 6 in the Core clauses.

- The Project Manager may instruct the Contractor to submit a quotation to deal with the monetary losses associated with the delays (62).
- Assessment of Compensation Events is a matter for the Project Manager in accordance with certain rules and procedures (63 and 64). The Project Manager must notify the Contractor of his or her decision.

3 Control

- The early warning obligation can constitute a control mechanism (16).
- The Project Manager and the Supervisor are both given considerable powers, and both may delegate. The Project Manager is empowered to issue instructions to the Contractor relating to changes of Works Information (14).
- The Contractor must submit names, qualifications and experience of key people. Replacements are subject to acceptance by the Project Manager (24).
- The Project Manager may order the removal of an employee from any further connection with the particular contract (24).
- The Contractor must arrange for access to Works, materials and plant in store, for the Project Manager, Supervisor and others notified by the Project Manager (28).
- The Contractor must obey instructions from the Project Manager and the Supervisor, authorised under the contract (29).
- The Contractor is wholly responsible for the work of all sub-contractors. All sub-contractors are subject to acceptance by the Project Manager (26).
- Tests and inspections carried out by the Contractor may be observed by the Supervisor, who may also order the Contractor to carry out tests, and carry out his own tests (40).
- The Supervisor may instruct the Contractor to search for defects (42) and whether or not the Supervisor notifies him the Contractor is obliged to correct defects before the end of the Defects Correction Period (43).
- Where Works have been taken over by the Employer, the Project Manager is to arrange access for the Contractor to correct defects. Where it is agreed that defects need not be corrected, then there may be changes to Works Information and a cost reduction (44).
- The contract makes no reference to assignment, other than the Contractor's obligation to assign the benefit of contracts on termination of his employment (96).

NEC

4 Money

- An Addendum Y(UK)2 (essential for all contracts on ECC in the United Kingdom), takes account of the Housing Grants, Construction and Regeneration Act 1996 and this affects the provisions for payment as set out in Section 5 of the Core clauses.
- The Project Manager must assess the amounts due to the Contractor at each assessment date. The intervals are included in the Contract Data (50).
- The Project Manager certifies a payment on or before the date on which a payment becomes due, and each certified payment is made on or before the final date for payment (51).
- The Project Manager's certificate must show the amount of payment due, subject to any Employer's rights of deduction, and the basis on which the figures are calculated (56, incorporated by Y(UK)2).
- The date on which payment becomes due is seven days after the assessment date, and the final date for payment is 21 days after the date on which payment becomes due (56, incorporated by Y(UK)2).
- If the Employer intends to withhold payment after the final date, he must notify the Contractor at least seven days before the final date, stating the amount he intends to withhold and the grounds for so doing (56, incorporated by Y(UK)2).
- Interest is payable on amounts due but unpaid, and in respect of a Project Manager's certificate which is due but issued late. The interest rates are to be stated in the Contract Data (51).
- Suspension of performance by the Contractor in the event of the Employer's failure to make proper payment is treated as a Compensation Event (60·7, incorporated by Y(UK)2).
- There can be price adjustments for inflation (i.e. fluctuations) under Supplementary Option N (applicable only for Main Options A–D) calculated on the basis of a Price Adjustment Factor.

5 Statutory obligations

- The Contractor is obliged to notify the Project Manager if he becomes aware of anything in the Works Information which would be illegal (19).
- The Contractor is contractually bound to act in accordance with health and safety requirements, and work additional to that included in Works Information may be a Compensation Event (18). (See also Secondary Option clause U.)

6 Insurance

- The standard risks to be carried by the Employer are itemised, and any additional risks to be carried by the Employer may be entered in the Contract Data (80).

- Risks not itemised as being carried by the Employer are to be carried by the Contractor (81).
- Each party indemnifies the other in respect of claims, proceedings, compensation or costs arising from an event which is at risk of the party concerned (83).
- Insurance responsibilities are tabled and as stated in the Contract Data. Policies are to be taken out in joint names, and in the case of the Contractor policies and certificates are subject to acceptance by the Project Manager. If the Contractor fails to submit, then the Employer may insure and charge to the Contractor (85 and 86).
- Where the Employer insures, policies and certificates must be submitted to the Contractor. If the Employer fails to submit, then the Contractor may insure and charge to the Employer (87).
- Details of insurance obligations and cover are to be stated in the Contract Data.

7 Termination

- Valid reasons for termination by the Employer and by the Contractor are set out in a Termination Table (94).
- Either party is also given the right of termination by reason of insolvency or stated default by the other, or because of some action by the Project Manager (95).
- In the event of termination the Project Manager issues a Termination Certificate and within 13 weeks certifies final payment (94).
- After issue of the Termination Certificate, the Contractor is not obliged to carry out further work, and the Employer may order the Contractor to leave the site and remove the plant and materials.
- The Employer may elect to complete the Works himself or employ another contractor, and may use any plant or materials to which he has title (96 and 70) .
- Payments which may be due on termination are as set out in the Termination Table (97).

8 Miscellaneous

- Main Option clauses can include Activity Schedules (for A or C), or bills of quantities (for B or D).
- Use of the NEC engineering and construction sub-contract and the professional services contract seem to be required regardless of which Main Options are incorporated.
- Secondary Option clauses can be incorporated to make provision for performance bond (G); parent company guarantee (H); advanced payment to the Contractor (J);

multiple currencies (K); sectional completion (L); limitation of Contractor's design liability to using reasonable skill and care (M); price adjustment for inflation (N); retention money (P); bonus for early completion (Q); delay damages (R); low performance damages (S); changes in the law (T); CDM Regulations (U); trust fund (V); and additional conditions of contract (Z). For the latter to be incorporated, they should be stated in the Contract Data.

Option Y(UK)3 should be incorporated into the Contract Data for all contracts to which the law of England and Wales, and Northern Ireland applies, and is a contracting out of third party rights under the Contracts (Rights of Third Parties Act) 1999.

9 Disputes

- The Addendum Y(UK)2 affects provisions for adjudication as set out in Section 9 of the Core clauses.
- The Adjudication Table which appears in the Second Edition of the ECC together with the other provisions in clause 90 are deleted, and replaced by a new clause 90 'Avoidance and settlement of disputes'. Under this the parties to the contract and the Project Manager are to follow the detailed procedures as set out.
- The adjudicator may be named in the Contract Data.
- The adjudicator's decision is final and binding unless and until referred to a further 'tribunal'. Whether this is to be arbitration or legal proceedings will presumably be stated in the Contract Data. If arbitration, the procedures to be followed appear not to be stated in the contract and would presumably be for the parties to agree.

NEC

This contract?

If considering using the ECC remember that:

The idea of Core clauses to which selected Optional clauses and Secondary Option clauses may be added to produce a contract which can be tailored to a particular set of circumstances is seductively attractive. There are risks in this approach, however, and the various options can result in conditions which some lawyers regard as producing great variability and a risk of imbalance.

The Contract Data and Works Information are essential components of the ECC. The former is not to be changed once the contract is entered into. The Works Information supplied by the Employer and by the Contractor can be presented in a variety of formats, and is information necessary at tender stage. Additionally Addendum Y(UK)2 takes account of the Housing Grants, Construction and Regeneration Act 1996, and Addendum Y(UK)3 takes account of the Contracts (Rights of Third Parties) Act 1999.

There are no stated restrictions on the use of the ECC, but the law of the contract, the language of the contract, and the currency of the contract should be entered in Part One of the Contract Data. Changes in the law of the country in which the site is located might be a Compensation Event if Option T is incorporated. Addendum Y(UK)2 will be applicable for England and Wales, Northern Ireland and Scotland, and Y(UK)3 will be applicable only to England and Wales and Northern Ireland.

As might be expected in a form with this engineering provenance, there is no reference to an architect or quantity surveyor by profession. The key persons are the Project Manager, who manages the procurement of the Works for the Employer; and the Supervisor, who exercises certain responsibilities on site for the Employer with whom he or she has a contract for services.

Completion of the form requires selecting or assembling the appropriate options, and making relevant entries to 11 pages of Contract Data. Attestation is by a separate document.

Contract administration requires attention to communications, early warnings, changes in Works Information, and Compensation Events in particular. A cooperative and non-adversarial attitude is essential with ECC, and there is a strong emphasis on best practice management.

There is as yet very little legal pronouncement, but some well respected legal commentators initially expressed reservation over what they consider to be many unresolved legal issues. However, the form seems to have been widely used without serious problems.

6 Traditional procurement: standard lump sum forms

NEC

Related matters

Documents

Engineering and Construction Contract (ECC) Second Edition (November 1995)
Addendum Y(UK)2 (April 1998)
Addendum Y(UK)3 (April 2000)
NEC Partnering Option, Option X12 (June 2001)

Note: The Engineering and Construction Contract is available as an Omnibus Edition (Black Book) which contains the Core clauses and all Optional clauses. The ECC is also available as separate documents which contain the Core clauses merged with the applicable main Option Clauses:

ECC Option A. Priced contract with Activity Schedule
ECC Option B. Priced contract with bill of quantities
ECC Option C. Target contract with Activity Schedule
ECC Option D. Target contract with bill of quantities
ECC Option E. Cost reimbursable contract
ECC Option F. Management contract
Engineering and Construction Sub-contract
Guidance Notes for the Engineering and Construction Contract
Flow Charts for the Engineering and Construction Contract

Professional Services Contract
Adjudicator's Contract

Commentary (relating to ECC)

Brian Eggleston
The New Engineering Contract: A Commentary
Blackwell Science (2000)

Traditional procurement 7

Shorter lump sum forms

The Joint Contracts Tribunal Ltd
Intermediate Form of Building Contract for Works of Simple Content
1998 Edition

The Joint Contracts Tribunal Ltd
Agreement for Minor Building Works 1998 Edition

Association of Consultant Architects
ACA Form of Building Agreement 1982 (Third Edition 1998, 2000 Revision)

The Stationery Office
Form GC/Works/2 (1998)

The Stationery Office
Form GC/Works/4 (1998)

The Institution of Civil Engineers
NEC Document
Engineering and Construction Short Contract (July 1999)

The Joint Council for Landscape Industries (JCLI)
Agreement for Landscape Works

Architecture and Surveying Institute
ASI Forms of Contract

Scottish Building Contract Committee (SBCC)
SBCC Forms of Contract

Just what constitutes a shorter form is arguable. For the purposes of this book, contracts which are comparatively brief and easy to handle in terms of administration are included under this heading. This need not necessarily imply that they are solely for smaller works nor indeed only suitable for lump sum contracts.

See also Chapter 8 below, in which short forms which are more likely to be categorised as consumer contracts are covered.

IFC98

The Joint Contracts Tribunal Ltd

Intermediate Form of Building Contract for Works of Simple Content 1998 Edition (IFC98)

Background

With the introduction of a then new and more sophisticated edition of The Standard Form of Building Contract (JCT80), many architects felt that they lacked a contract suitable for middle range jobs. The RIBA expressed this concern, and in October 1981 the JCT set up a working party to prepare an 'intermediate' form. With considerable input from constituent bodies, in particular the RIBA, the Association of District Councils and the Association of Metropolitan Authorities, the Intermediate Form appeared in September 1984.

IFC84 was the subject of 12 Amendments, the last of which was to meet many of the recommendations made in the Latham Report of 1994. Incorporation of Amendment 12 ensured compliance with Part II of the Housing Grants, Construction and Regeneration Act 1996 in respect of adjudication and payment provisions.

The 1998 Edition of the Intermediate Form is basically a consolidated version of IFC84 but since publication has received four Amendments.

Nature

The form now runs to over 60 pages. There are five Recitals (the Fourth of which includes an Information Release Schedule) and nine Articles. The Conditions are arranged under the now customary section headings. There is a contents list at the front which gives clause and page numbers. In addition to an Appendix, the form includes Supplemental Conditions in respect of VAT, Construction Industry Scheme, and reference to the Fluctuations Conditions which are published in a separate booklet. There is also an Annex to the Appendix concerning bonds, and an Annex to the Conditions concerning EDI.

IFC98 has the virtue of relative brevity, a clear layout and commendably easy cross-referencing. Although the wording of the clauses is of necessity truncated in parts, the Conditions should be adequate for the foreseeable circumstances of most middle range projects.

The Intermediate Form can be modified by the use of a Sectional Completion Supplement (Article 7) and the Fluctuations Supplement, but there is no supplement to cover design by the Contractor and no provision for performance specification.

7 Traditional procurement: shorter lump sum forms

IFC98

Use

The rear cover of the form (not part of the text) lists three criteria for suitability. These refer to building works of simple content and without complex services installations. Most important, there is a reminder that this is a lump sum contract, to be priced entirely by the Contractor, and therefore the job must be fully designed and billed or specified at tender stage. JCT Practice Note 5 (Series 2) repeats these points and suggests that suitability is indicated by a contract value of up to £375,000 (at 2001 prices), and a contract period not exceeding 12 months.

Practice Note 5 accepts that contracts may well be larger or longer than the limits suggested. Experience suggests that the form can be used successfully on considerably larger projects and it is the nature of the work which should be the ultimate determinant.

There is one version of the form only, to cover use in private or local authority sectors, and it can be used with or without quantities. There is no version for approximate quantities.

The Employer is required to appoint an Architect/Contract Administrator (Article 3), and to name whoever is appointed to undertake the duties required of a quantity surveyor (Article 4). The Planning Supervisor and Principal Contractor will be identified in entries to Articles 5 and 6.

The First Recital identifies the contract documents as being the Contract Drawings, and either the bills of quantities or the Specification/Schedules of Work. Where a sub-contractor is named, the NAM documents are also to be included.

The Second Recital allows the Contractor to tender either by pricing the itemised bills, Specification or Schedules, or to state just a lump sum based on a Specification which is not itemised for pricing. If the latter then the Employer will require the contractor to submit a Schedule of Rates, or a Contract Sum Analysis, or a priced Activity Schedule. This will not be a contract document, but it is an important supporting price document, and the Employer might be wise to stipulate a Contract Sum Analysis option. He might also require it to be in a preferred format, and perhaps this should be prepared by the quantity surveyor for completion by the tenderers.

There is a Sectional Completion Supplement for IFC98, and if partial possession before Practical Completion is required, then clause 2·11 provides for this.

Synopsis

1 Intentions

- The Contractor is obliged to carry out and complete the Works in accordance with the contract documents and with the Health and Safety Plan (1·1).
- The quality and quantity of work included in the Contract Sum is clearly defined (1·2)

and is related to the documents used.

- The contract documents are to be read as a whole (8·2) and the printed Articles, Conditions and Appendix have priority (1·3).
- In the case of inconsistencies or errors in or between documents (including the tender particulars for a named sub-contractor) corrective instructions must be given (1·4).
- Contract bills, except where specifically stated otherwise, must be prepared in accordance with the Standard Method of Measurement 7th Edition (1·5).
- If applicable, the Contractor must be provided with two copies of the information referred to in the Information Release Schedule by the stated times (Fourth Recital and clause 1·7). The Contractor must also be issued with further drawings as reasonably necessary to complete the Works (1·7). Use of them is limited and confidentiality of rates is to be respected (1·8).
- There is no reference to a Contractor's programme.
- The CDM Regulations oblige the Employer to nominate a Planning Supervisor and where relevant a Principal Contractor. This becomes a contractual obligation also (Articles 5 and 6 and clause 5·7) and any need to appoint a replacement is also covered (1·12).

2 Time

- Dates for possession and completion must be entered in the Appendix. The Contractor must proceed regularly and diligently and complete on or before the completion date (2·1).
- An option clause for deferring possession not exceeding six weeks may apply subject to an Appendix entry (2·2).
- Notice of delay must be given in writing by the Contractor, and he must supply any information reasonably necessary. The Architect is to consider a new completion date and to notify the Contractor of his or her decision, in writing, as soon as he or she is able to see the effect on completion. Although no time limit is stated, this must be within a reasonable time (2·3). 'Events' are listed for which the Architect is empowered to make an extension (2·4). Review of extensions up to 12 weeks beyond Practical Completion is discretionary (2·3).
- If the Contractor fails to meet the completion date, this fact must be certified (2·6). Liquidated damages may be deducted or otherwise recovered by the Employer (2·7).
- Practical completion is certified by the Architect (2·9).
- The Contractor is obliged to rectify defects (2·10) unless the Employer decides otherwise and takes an appropriate deduction instead.

IFC98

- There is provision for partial possession (2·11).
- Where the Employer wishes to occupy part of the uncompleted Works for storage etc., this is possible subject to proper insurance arrangements (2·1).

3 Control

- The bar to assignment without written consent refers to 'this contract' (3·1). Consent to sub-contract any part of the work requires the Architect's written consent (3·2).
- All sub-contractors are domestic. There is provision for naming a single sub-contractor (3·3). Two methods are available, but both require the use of a tender document NAM/T and sub-contract conditions NAM/SC.
- There is no provision for a list of three names, and such a practice should not be necessary. However, the NBS Small Jobs Version does offer suitable clauses, although care should be taken to state that this is a 3·2 sub-contract and not under clause 3·3.
- Architect's instructions must be in writing (3·5). Instructions empowered include for variations (3·6), expenditure of provisional sums (3·8) and postponement (3·15).
- The Contractor is required to have a competent person-in-charge on the Works at all reasonable times (3·4) and to permit the presence of the Employer's Clerk of Works (3·10) who has no prescribed authority.
- Where work or materials do not conform to the contract, the Architect may order their removal from the site (3·14). Instructions are empowered concerning inspection and testing (3·12) and there is a particularly helpful provision concerning similar work elsewhere which may be suspect following established failure (3·13).
- The Employer is entitled to carry out work not forming part of the contract during the time that the Contractor is in possession (3·11).

4 Money

- Mainly because of the Latham Report and consequential legislation, these conditions and procedures are quite stringent. They are an attempt to aid cash flow, bring interest to sums not paid when due, and provide valid grounds for suspending work in the event of non-payment.
- The Contract Sum is VAT-exclusive (Article 2) and may only be adjusted as provided for in the contract (4·1).
- The contract may include for fluctuations (4·9). These may be limited to tax etc. (Supplemental Condition C) or, only where bills of quantities apply, they can be by application of formula rules (Supplemental Condition D).
- Where provisional sums have been included, instructions must be given (3·8). Caution is needed, if SMM7 applies, to distinguish between defined and undefined work and

this is particularly important where a provisional sum has been included for named sub-contract work (3·3·2).

- Applications for reimbursement of loss and/or expense must be made in writing by the Contractor (4·11). The grounds for a valid application are set out (4·12) and include only matters over which the Contractor has no control and which occur because of action or failure by the Employer. The procedures must be followed precisely. Other rights at common law are preserved.
- Interim payments to the Contractor can be at pre-arranged stages (4·2) or at intervals of one month (unless stated otherwise). Retention is 5 per cent, but is not referred to by that name.
- Interim Certificates show the amount due and state the basis of calculations. Within five days of issue, the Employer must notify the Contractor in writing of the amount of payment he intends to make. Any intention to withhold or deduct must be clearly stated in a written notice which also sets out the grounds for such action, and this must be given to the Contractor not less than five days before the date for final payment. The final date for payment is 14 days from the date of issue of the Interim Certificate, and if no valid notices have been served by the Employer payment in full is required. Failure to pay by the final date for payment will attract interest of 5 per cent over current base rate, and can give the Contractor a right to suspend work (4·2).
- Within 14 days after Practical Completion, an Interim Certificate should be issued, after which only half the retention money is held (4·3).
- There is a precise timescale for the preparation of a final account after Practical Completion (4·5). Issue of the Final Certificate is within 28 days of sending the final account to the Contractor, or certifying that defects have been made good, whichever is the later (4·5 and 4·6).
- Similar rules apply in respect of the Final Certificate with regard to notices, and the final date for payment is 28 days from the date of issue of the Final Certificate (4·6).

5 Statutory obligations

- It is the Contractor's duty to comply with all statutory obligations and give all required notices (5·1). He is entitled to recover fees and charges not otherwise provided for (5·1).
- The Contractor is to notify the Architect if he finds any conflict between statutory requirements and the contract documents (5·2). The Architect must issue an instruction. The Contractor is then not liable for any non-compliance if it results from that instruction (5·3).
- The Contractor is empowered to carry out limited work for emergency compliance and this will be treated as a variation to be valued accordingly (5·4).

IFC98

- Where the CDM Regulations apply, the Contractor is contractually obliged to comply with all the relevant CDM duties, particularly with regard to the Health and Safety Plan, and in providing information for the Health and Safety File (5·7).

6 Insurance

- What a particular contract includes will depend to a large extent on the Appendix entries (e.g. whether option clauses are to apply, minimum amount of cover required, etc.). The Architect may be obliged to issue instructions, call for documentary evidence and pass policies to the Employer for checking.
- The Contractor is to indemnify the Employer in respect of personal injury or death, and injury or damage to property other than the actual Works (6·1 and 6·2). This is to be backed by insurance, and an Appendix entry will state the minimum cover.
- If instructed, the Contractor it to take out joint names insurance for the Employer against risk of claims arising due to legal nuisance. There is a list of exceptions, and damage must not be attributable to any negligence by the Contractor. An Appendix entry will indicate whether cover may be required and the amount of cover to be provided (6·2·4).
- Insurance of 'the Works' is for all risks where new buildings are concerned and should be for full reinstatement value. It can be taken out either by the Contractor (6·3A) or by the Employer (6·3B).
- Insurance of existing structures and the contents is a matter for the Employer (6·3C) and is limited to specified perils. New work in existing buildings, although still a matter for the Employer, requires all risks cover.
- Insurance may be required against the Employer's loss of liquidated damages due to an extension of time following damage to the Works (6·3D). This requires an Appendix entry and instructions to the Contractor.
- An Appendix entry will show whether the Joint Code of Practice on the Protection from Fire of Construction Sites is to apply (6·3FC) and if so, both the Employer and the Contractor must comply with it. In the event of non-compliance the insurers can specify remedial measures which have to be taken.
- In the event that terrorism cover is withdrawn and is no longer available, the situation and options open to the Employer are dealt with in clauses 6·3A·5, 6·3B·4 or 6·3C·5 as applicable.

7 Termination

- The Employer is allowed to determine the employment of the Contractor by reason of specified defaults (7·2). A warning notice is required before the actual notice of determination by the Employer.

IFC98

- In the event of the Contractor's insolvency, then depending on the circumstances, determination might be automatic but subject to possible reinstatement, or the Employer might enter into an agreement with the Contractor (a '7·5·2·1 Agreement') for continuation or novation (7·5).
- Depending on the circumstances the Employer can have the right to make interim arrangements for certain works to be carried out during the time that the works are on hold under a 7·5·2·1 Agreement, or to have the works completed by another contractor, or to decide against having the works completed at all (7·7).
- The Contractor is allowed to determine his own employment for specified defaults of the Employer (7·9). In the event of the insolvency of the Employer, the Contractor may elect to determine his own employment (7·10).
- Either party can determine the employment of the Contractor for listed neutral causes (7·13).
- The respective rights and duties of the parties concerning payment, removal and completion are set out in detail (7·6, 7·11 or 7·14).

8 Miscellaneous

- Definitions are included (8·3).
- There is a contracting out of third party rights under the Contracts (Rights of Third Parties) Act 1999 (1.17).
- There is no reference to access for the Architect, but this is probably implied.
- There is no power to exclude persons.
- There is no provision for antiquities.
- Contract Sum Analysis (Second Recital) is defined (8·3), and JCT Practice Note 23 (original series, green cover) gives a useful explanation.
- The use of NAM/T where a person is to be a named sub-contractor is confirmed (3·3).

9 Disputes

- Part II of the Housing Grants, Construction and Regeneration Act 1996 gives either party a statutory right to refer any difference or dispute arising out of the contract to adjudication. Article 8 of IFC98 provides for this.
- Procedures for referral to adjudication, the appointment of an adjudicator, the powers and conduct of an adjudicator are as set out (9A).
- The adjudicator's decision is binding on the parties at least until the dispute is finally determined at arbitration or by legal proceedings.

IFC98

- Article 9A establishes arbitration as the agreed method for final determination of disputes (9B).
- The appointment of the arbitrator is subject to clause 9B·1, and the arbitrator's powers are clearly defined.
- The parties agree that either may appeal to the courts on a question of law (9B·4).
- Arbitration is to be conducted in accordance with the JCT 1998 Edition of the Construction Industry Model Arbitration Rules and the provisions of the Arbitration Act 1996 shall apply.
- Where the Appendix shows that clause 9B is deleted, then clause 9C, legal proceedings, will automatically apply (see also Article 9B).

IFC98

This contract?

If considering using IFC98 remember that:

It is intended for building work of a simple content, without specialist installations, and where the work is adequately specified or billed before tender. It is in one version only, with or without quantities, and for use in either private or public sectors. The Employer is required to appoint a Contract Administrator and a quantity surveyor.

If used for work in Northern Ireland an Adaptation Schedule should be incorporated. The form is not suitable for use in Scotland, and where the site of the works is in Scotland, the appropriate SBC form should be used.

It can include for partial possession and sectional completion, but there is no provision for design by the Contractor or performance specified work.

It allows for sub-contractors to be named, and there are two procedures. The first requires the naming to be pre-contract and may seem inflexible but brings greater certainty. The second allows for naming during construction and is covered in the contract by a provisional sum. In both cases the sub-contractors are domestic and the responsibility of the main Contractor. Use of dedicated documents is mandatory with named sub-contractors. As the main Contractor is not responsible for design, employers should be advised to use an ESA/1 agreement for each named sub-contractor involved in some design relating to their work.

When completing the form, decisions are required relating to matters including deferment of possession; bonds (for advance payment, 'listed items'); insurance of the works; Joint Fire Code; liquidated damages; fluctuations; and EDI. Caution is needed over naming a quantity surveyor – this should not be left blank, and even if the Employer does not agree to appointing a quantity surveyor, the mechanism of the contract still requires a name to be entered to fulfil this role (usually that of the Architect).

If acting as Contract Administrator note that while the procedural rules are less demanding than those for JCT98, care is needed to make sure that action is taken within a reasonable time even though no time limit is stated. This is particularly so concerning named sub-contractors.

Amendments are issued by the JCT from time to time. The form is available on disk. The RIBA publishes contract administration forms for IFC98.

The JCT suggests that IFC98 is appropriate for contracts up to £375,000 (2001 prices) in value, but experience indicates that it is commonly used in excess of this figure. The form is logically structured with section headed conditions, and written in clear English. The lack of provision for design responsibility by the main Contractor is perhaps a regrettable omission. However, no real difficulties seem to have arisen to date, and the form is deservedly popular.

IFC98

Related matters

Documents

Intermediate Form of Building Contract IFC98
Amendment 1: 1999 (Construction Industry Scheme)
Amendment 2: 2000 (sundry amendments)
Amendment 3: 2001 (terrorism/Joint Fire Code/SMM)
Amendment 4: 2002 (extension of time/loss and expense/advance payment)

Supplements and ancillary:
IFC/FS Fluctuations Supplement
IFC/SCS Sectional Completion Supplement
NAM/T98 Tender and Agreement (named sub-contractors)
NAM/SC98 Sub-Contract Conditions (named sub-contractors)
NAM/FR98 Formula Rules (named sub-contractors)
ESA/1 Employer/Specialist Agreement 2001 Edition (RIBA/Specialist Engineering Group publication)

References

Series 2 JCT Practice Notes (yellow covers):
Practice Note 1: Construction Industry Scheme
Practice Note 2: Adjudication (includes text of agreements)
Practice Note 3: Insurance, Terrorism Cover
Practice Note 4: Partnering (includes text of non-binding charter)
Practice Note 5: Deciding on the Appropriate JCT Form of Contract
Practice Note 6: Main Contract Tendering (includes model forms)
JCT Guidance Note for IFC98

Commentaries (relating to IFC98)

David Chappell and Vincent Powell-Smith
The JCT Intermediate Form of Contract
2nd edn, Blackwell Science (1999)

Sarah Lupton
Guide to IFC98
RIBA Publications (2001)

Neil Jones and Simon Bayliss
Jones and Bergman's JCT Intermediate Form of Contract
3rd edn, Blackwell Science (1999)

RIBA Publications
Architect's Guide to the Contract Administration Forms IFC98
RIBA Publications (2000)

MW98

The Joint Contracts Tribunal Ltd

Agreement for Minor Building Works 1998 Edition

Background

The Agreement for Minor Building Works first appeared in 1968, and was intended for minor building operations and maintenance work for which the JCT Standard Form of Building Contract (then JCT63) was clearly inappropriate. It was just five pages long, compared with nearly 40 pages of the full Standard Form. It was not for use with bills of quantities, and was stated as being suitable where the Contract Sum did not exceed £8,000.

In the 1970s the RIBA, conscious of certain deficiencies in both the 1963 Standard Form and the Minor Works Agreement, set out proposals for a new, simpler 'short form' of building contract. RIBA Council approved this proposal with acclamation in June 1978. The aim was for a four page contract with conditions written in plain English and structured logically under eight headings. The JCT accepted the proposal in principle, but the response of their working group was to draft a new 1980 version of the Minor Works form. It was a significant advance, however, and this was the first time that Conditions in JCT Forms were arranged under section headings. Incidentally the 'short form' concept was later to resurface in the drafting of IFC84.

MW80 was the subject of 11 Amendments, the last of which was to take account of the Housing Grants, Construction and Regeneration Act 1996 (Part II). The 1998 Edition of the Agreement for Minor Building Works is basically a consolidated version of MW80 but since publication has received four Amendments.

In many instances where the Minor Works Agreement is used for domestic work, this legislation will not apply and the contract will not be subject to the statutory requirements relating to adjudication and payment provisions. Nevertheless, MW98 includes for adjudication and certain procedures relating to payment because even if not required by statute, these are generally regarded as being good practice. They should be removed only for good reason and after taking legal advice.

The current edition of the form carries a warning that it is only intended for use where the client has engaged a professional consultant to advise on and to administer the terms. It may not be suitable for use as a contract direct between consumer and contractor of the type to which the Unfair Terms in Consumer Contracts Regulations 1994 (SI 1994/3159) might apply. The JCT now has other contracts specially drafted for use by consumers in connection with work on their homes, whether or not a consultant has been engaged.

7 Traditional procurement: shorter lump sum forms

MW98

Nature

The total document is over 30 pages long, but the contract Conditions take up only eight of these. The Agreement includes five Recitals and seven Articles, and the Conditions are set out under section headings. There is no appendix, but entries are to be made in the text of the Conditions.

Supplemental Conditions are printed as part of the document. Condition A on contribution, levy and tax changes is incorporated by clause 4·6. Condition B is the VAT Agreement and is incorporated by clause 5·2. Condition C is the Construction Industry Scheme, incorporated by clause 5·3. Condition D is Adjudication, incorporated by Article 6 and clause 8·1, and Condition E is Arbitration, incorporated by Article 7A and clause 8.2.

There is one version only of MW98 for use in both private and public sector.

Use

The form is relatively brief, and apparently simple. However, it should be remembered that what is expressly stated might not be the entire picture. Terms might be implied by common law, and the form needs to be used with thought and treated with care.

A Guidance Note on the use of MW98 is included with the form, and gives a clear reminder that it is to be used only where the client has engaged a professionally qualified person to act as the Architect/the Contract Administrator in administering the terms. It is for use where minor building works of simple character are to be carried out for an agreed lump sum. The form is not for use with works for which bills of quantities are required, or where the duration is likely to be such that full labour and materials fluctuations provisions are required. There is no provision in the form for the Employer to nominate or name sub-contractors for specialist work, and no dedicated domestic sub-contract is published.

There can be 'contract drawings', a 'contract specification' or Schedules which are also contract documents (First Recital). The Contractor will price a detailed contract document, or may provide a lump sum price supported by a Schedule of Rates (Second Recital). There are no supplements published to extend the range of this contract.

JCT Series 2 Practice Note 5 states that MW98 is suitable for contracts up to the value of £100,000 at 2001 prices.

Synopsis

1 Intentions

- The Contractor is obliged to carry out and complete the Works in accordance with the contract documents and the Health and Safety Plan where applicable (1·1).

MW98

- The Contractor is to perform with due diligence and in a good and workmanlike manner (1·1).
- The quality and standards are those described in the contract documents (1·1).
- The Architect is obliged to issue any further information necessary to enable the Contractor to carry out the Works (1·2).
- Inconsistencies in or between contract documents shall be corrected and this may be treated as a variation instruction (3·6).
- The printed Conditions prevail (3·6).
- Where all the CDM Regulations apply a Planning Supervisor and Principal Contractor must be identified, and a reappointment made if necessary (Fifth Recital, Articles 4 and 5 and clause 1·3). Where the work is such that CDM Regulations 7 and 13 only apply, the Contractor is to notify the Health and Safety Executive (1·4).

2 Time

- The Works may be commenced on a date to be inserted and shall be completed by a date to be inserted (2·1).
- The Contractor is to notify the Architect if completion by the stated date is unlikely, for reasons which are not within the Contractor's control (2·2).
- The Architect is empowered to make such extension of time as may be reasonable. (Note: no specific grounds are listed).
- If the Contractor fails to complete by the date, the Employer is entitled to liquidated damages (2·3) and the rate is to be inserted. (Note: there is no reference to a certificate of non-completion.)
- Practical Completion is certified by the Architect (2·4).
- Following this the Contractor is obliged to rectify defects. (The Defects Liability Period is three months, although a longer period may be inserted, 2·5). A certificate is to be issued by the Architect when this obligation has been discharged.
- There is no provision for deferring possession, nor for partial possession, nor for sectional completion.

3 Control

- The bar to assignment without written consent refers to 'this Agreement' (3·1). Written consent to sub-contract is required (3·2).
- There is no provision for naming a sub-contractor, and no provision for including a list of approved firms. Note however that neither is there anything which prevents this, and the NBS Small Jobs Version suggest suitable ways of achieving this.

MW98

- Architect's instructions are to be issued in writing (3·5) but may be oral and confirmed in writing. Instructions may include variations (3·7) and this would presumably allow for postponement, as no specific reference is made elsewhere.
- The Contractor's 'competent person' is to be on site at all reasonable times (not constantly) (3·3). There is no provision for a Clerk of Works.
- There is no provision for testing, or opening up of work. In the event of failure to meet the contract standards, and problems over remedial works, refusal to certify leading to non-payment would seem to be the ultimate sanction.
- There is no provision for work to be carried out by persons engaged directly by the Employer.

4 Money

- The Contract Sum is VAT-exclusive (Article 2).
- The contract is fixed price (4·7) except for limited fluctuations (tax etc.) provided for by Supplemental Condition A (4·6). A percentage addition entry is required. (If none, NIL is entered).
- Instructions must be given if provisional sums are included (3·8). The cost of such work is either to be agreed in advance or valued (3·7).
- Direct loss and/or expense is limited to variations or provisional sum work, and included in the valuation (3·7).
- Progress payments are to be made at four-weekly intervals (4·2). The Architect is to issue certificates (presumably to the Employer). Retention will be at 5 per cent (unless a different figure is entered).
- Interim Certificates will show the amount due and the basis of calculations. The Employer must notify the Contractor in writing of the amount he intends to pay not later than five days after the issue of the certificate. Any intention to withhold or deduct sums must be given by written notice and the grounds for such intention must be clearly stated. This notice must be given to the Contractor no less than five days before the final date for payment. The final date for payment is 14 days from the date of issue of the Interim Certificate, and if no valid notices have been given, then payment in full is required. Failure to pay by the final date for payment will attract interest of 5 per cent over the current base rate, and can also give the Contractor a valid right to stop work (4·8).
- A penultimate certificate is to be issued within 14 days after Practical Completion releasing all outstanding money except half the retention (4·3).
- The Final Certificate is to be issued within 28 days of either issue of the certificate signifying completion of making good the defects, or the Contractor submitting all

necessary information for the final account (4·5). He is given three months (unless a different figure is inserted) from the date of Practical Completion to produce this.

- Similar rules apply to the Final Certificate with regard to notices as those for Interim Certificates. The final date for payment in this case is stated as being 14 days from the date of issue of the Final Certificate.

5 Statutory obligations

- It is the Contractor's duty to comply with all statutory obligations, including giving all required notices (5·1).
- The Contractor is to notify the Architect in writing if he finds any conflict between statutory requirements and his other contractual obligations (5·1). Having done that, he will not be liable to the Employer under this contract.
- Where the CDM Regulations apply in full, the Contractor is contractually obliged to comply with all the relevant CDM duties, particularly in respect of the Health and Safety Plan and the Health and Safety File (5·7 and 5·9).

6 Insurance

- The Contractor indemnifies the Employer in respect of personal injury or damage to property. He is required to arrange insurance to back this (6·1 and 6·2). The minimum cover as a contractual obligation requires an entry (6·2).
- The Contractor is obliged to arrange insurance of new works in joint names against damage by perils specified in clause 6·3A. The percentage to cover professional fees should be inserted.
- There is no provision for the Employer to insure new works.
- Joint names insurance of existing structures, and any new work which is part of an alteration or conversion, is to be arranged by the Employer. The contractual requirement is for cover against specified perils (6·3B).

7 Termination

- The Employer may determine the employment of the Contractor if he fails to proceed satisfactorily with the Works, or if he becomes insolvent (7·2). Note that the latter does not mean automatic determination. The Architect may issue a warning notice, but notice of determination is a matter for the Employer.
- The Contractor may determine his own employment for stated reasons (7·3).
- There is no provision for determination by either party for neutral causes.
- The respective rights and duties of the parties following determination are set out in the Conditions (7·2 and 7·3).

MW98

8 Miscellaneous

- There is no definitions clause.
- There is no reference to access for the Architect, presumably this is implied.
- The Architect may order the exclusion of any person from the Works (3·4).
- There is no reference to antiquities, terrorism cover, etc.
- There is a contracting out of third party rights under the Contracts (Rights of Third Parties) Act 1999 (1·8).

9 Disputes

- Part II of the Housing Grants, Construction and Regeneration Act 1996 gives either party a statutory right to refer any difference or dispute arising out of the contract to adjudication. Article 6 of MW98 provides for this.
- Procedures for referral to adjudication, and the appointing of an adjudicator, the powers and conduct of an adjudicator are set out (8·1 and Supplemental Condition D).
- The adjudicator's decision is binding on the parties, at least until the dispute is finally determined at arbitration.
- Article 7A establishes arbitration as the agreed final method of settling disputes (8·2 and Supplemental Condition E).
- The appointment of the arbitrator is subject to Article 7A, and his or her powers are defined in Supplemental Condition E.
- Arbitration is to be conducted in accordance with the JCT 1998 Edition of the Construction Industry Model Arbitration Rules.
- If disputes are finally to be decided by legal proceedings (8·3 and Article 7B), Article 7A is deleted.

MW98

This contract?

If considering using MW98 remember that:

It is intended for small building work of a simple 'one-off' nature, and is not suitable for jobbing or maintenance type work. It is in one version only, and for use in either private or public sectors. It can only be used where the Employer has engaged a professional consultant to act as Contract Administrator. There is no provision for bills of quantities and although a quantity surveyor may be named, it is with no specific role under the Conditions.

If used for work in Northern Ireland an Adaptation Schedule should be incorporated. The form is not suitable for use in Scotland, and where the intended work is in Scotland, the SBC Scottish Minor Works form is available.

There is no provision for phased completion, naming of sub-contractors, fluctuations, or design by the contractor. The form is drafted to include compliance with the CDM Regulations, and with the Housing Grants, Construction and Regeneration Act 1996, whether or not these apply in full to the particular contract. When completing the form decisions are required relating to insurance of the Works and damages for non-completion. Note that there is no conventional appendix to the form, and entries are to be made in the text of the Conditions.

If acting as Contract Administrator, note that although the Conditions are likely to prove adequate for most situations, should the nature of the work require it, then it might be advisable to clarify the procedural rules pre-contract, especially if working with a Contractor for the first time.

Amendments are issued by the JCT from time to time. The form is available on disk. The RIBA publishes contract administration forms for MW98.

The JCT suggests that MW98 is appropriate for contracts up to £100,000 (2001 prices) in value, but it is sometimes used incautiously well beyond this. Although the conditions are clearly worded and the procedural rules are brief and easy to administer, it should be remembered that the law may imply terms in addition to those expressly in the contract. The form is perhaps more suited to commercial rather than domestic work where there can be problems with the client remaining in occupation. For slightly larger projects, or those which require more comprehensive conditions, then IFC98 might be a safer choice. For work on domestic projects, perhaps consumer contracts, then the new JCT forms for home owners might be more applicable. The Minor Works Agreement has featured in a surprising number of court cases, often due to use beyond its intended limits or because of careless administration, and a RIBA expert has commented that although it appears to be a favourite with the profession, at the same time it appears to be little understood. It has limitations, particularly concerning the insurance and determination provisions, and needs to be treated with respect and administered with diligence.

7 Traditional procurement: shorter lump sum forms

MW98

Related matters

Documents

Agreement for Minor Building Works 1998 Edition MW98
Amendment 1: 1999 (Construction Industry Scheme)
Amendment 2: 2000 (sundry amendments)
Amendment 3: 2001 (CIS Supplemental Condition C)
Amendment 4: 2002 (dispute resolution/defects liability/payment/attestation)

References

Series 2 JCT Practice Notes (yellow covers):
Practice Note 1: Construction Industry Scheme
Practice Note 2: Adjudication (includes text of agreements)
Practice Note 5: Deciding on the Appropriate JCT Form of Contract
Practice Note 6: Main Contract Tendering (includes model forms)
JCT Guidance Note for MW98 (included with form of contract)

Commentaries (relating to MW98)

David Chappell and Vincent Powell-Smith
The JCT Minor Works Form of Contract
2nd edn, Blackwell Science (1998)

Sarah Lupton
Guide to MW98
RIBA Publications (1999)

RIBA Publications
Architect's Guide to the Contract Administration Forms MW98
RIBA Publications (2000)

ACA Form

Association of Consultant Architects

ACA Form of Building Agreement 1982 (Third Edition 1998, 2000 Revision)

Background

When the ACA Form first appeared, it was seen by some as a reaction against the relatively complex procedures of the newly introduced JCT80. Two years later some of the advantages claimed for using the ACA Form evaporated when IFC84 was introduced. Nevertheless, the ACA Form was a new initiative; it is still an interesting form unencumbered by tradition, and has become firmly established. The first edition took two years and 11 drafts to produce. It was claimed that the new agreement was suitable for use on a very wide range of jobs of various sizes and types (although the initial reaction of legal commentators steeped in the phraseology of JCT documents was cautious, not to say dismissive).

At the outset the ACA contract working party claimed that their intention was to produce a form which was 'a flexible and fairly balanced contract for an efficient architect working with an efficient contractor for an efficient client'. Early criticism was that the form was drafted unilaterally without any contribution from contractors. Indeed the former National Federation of Building Trades Employers advised its members not to enter into contracts using the ACA Form. However, it has been used successfully on a wide variety of jobs, including fast-track projects. Major client bodies appear to have used the form without problems, and the British Property Federation cooperated over a version of the form suitable for use under the BPF System of procurement which was not withdrawn until 1998.

There is little evidence of problems with the form, and no body of case law has yet been built up. It may be that the ACA members using the form are, in line with the original intentions, efficient architects working with efficient contractors.

The second (1984) edition included a number of significant improvements. The form is kept under review, and revised from time to time. The Third Edition appeared in 1998, with amendments needed to take account of the Housing Grants, Construction and Regeneration Act 1996 (Part II).

The ACA has continued in its progressive and non-traditional path and recently launched the first standard partnering contract. (See Chapter 13 below for further details on PPC2000.)

Nature

The total document runs to over 30 pages. There is a contents page which lists the

ACA Form

sections in which the clauses of the Conditions are grouped. This is followed by the Agreement which includes alternative clauses relating to the Contract Administrator appointed; who will be responsible for the preparation of further drawings and details; and whether the Works are to be completed in sections.

Flexibility is a feature of the form, made possible by the use of alternatives both in the Agreement and also at various points throughout the Conditions. The parties execute the document either as a deed, or not as a deed, at the end of the Conditions. The expressed intention was to produce a form which was plainly worded. However, some of the terminology used is peculiar to the ACA form, and might cause confusion. The administrative procedures are relatively straightforward, although there is little clear guidance on their detailed use.

Examples of where the terminology differs from most other forms include:

- completion date: the alternative clauses for extensions of time, possible possession or taking-over in parts, and possible acceleration;
- extra costs: the production by the Contractor of his estimate of the cost implications of an Architect's instruction, for agreement, before complying with it;
- information: the attempt to establish systematically within the contract a schedule for the supply of information to be produced;
- design: the attempt to establish within the contract the extent to which Architect and Contractor are each liable for the design of the Works.

To help in making this feasible, there are Schedules which follow after the Conditions. The Time Schedule offers Alternative 1 where the job is to be completed as one operation, or Alternative 2 where the job is to be completed in sections. A further Schedule provides for detail on the issue of information; what this is to be, who is to be responsible, and when it is to be supplied. A final Schedule provides stage payment information.

Use

There is no suggested limit on cost or type of job for which the form is thought to be suitable. It can be used with or without a bill of quantities and where no quantity surveyor is appointed. It can be used where the employer is private or is a local authority, and where contract administration is by an Architect or by a supervising officer.

The contract can be fixed price, or fluctuations based on an ACA index can be incorporated. The form may be used as a traditional work and materials contract, or the contractor may be required to accept a measure of responsibility for design and the provision of drawings. The flexibility in the document is made possible only by the introduction of alternative clauses, and great care is needed to ensure that the intended alternatives are clearly evident.

ACA Form

The Third Edition takes account of the Housing Grants, Construction and Regeneration Act 1996 (Part II).

Synopsis

1 Intentions

- The Contractor is obliged to execute and complete the Works. This must be done in strict accordance with the contract documents (1·1). The Contractor is required to use skill, care and diligence (1·2).
- What constitutes 'the contract documents' will be defined in clause C of the Agreement, and incorporated by clause D.
- The provisions of the printed form will prevail, unless there is anything to the contrary such as documents listed under clause 1·3.
- The drawings may be supplied solely by the Architect (2·1 Alternative 1) or may be supplemented by drawings and details for which the Contractor takes responsibility (2·1 Alternative 2). This will also be evident under Alternative 1 or 2 in clause F of the Agreement.
- The Contractor is responsible for the accuracy of drawings and other information prepared by him, and warrants compliance with performance specifications and fitness for purpose (3·1).
- The Contractor is to permit others engaged directly by the Employer to carry out work which does not form part of the contract, whilst he has possession (10·1).

2 Time

- Time Schedule entries set out the important dates. Alternative 1 is used for a single phase job. Alternative 2 allows for possession in parts and/or completion by sections.
- The Employer is to give the Contractor possession to the date or dates stated in the Time Schedule, and the Contractor is to commence immediately and proceed regularly and diligently (11·1).
- The Contractor's entitlement to extensions of time will be subject to Alternative 1 or Alternative 2 of clause 11·5. The first alternative includes solely delay caused by the Employer, his Architect or persons appointed under the CDM Regulations. The second is wider, and includes for neutral causes. The Architect has 60 days in which to notify the Contractor of his or her decision, and decisions are subject to review within a reasonable time of taking over (11·6 and 11·7).
- The Architect is also empowered to issue an instruction to accelerate or postpone the dates set out in the Time Schedule. This is an unusual provision, although it may be subject to the test of reasonableness. There must of course be a fair and reasonable

ACA Form

adjustment to the Contract Sum as a result, and a revised Time Schedule is required (11·8 and 11·9).

- The onus is on the Contractor to notify the Architect when the Works are ready for taking over by the Employer. The Contractor may also produce a Contractor's List of outstanding items. The Architect may then issue a Taking Over Certificate, or issue an Architect's List of items which need attention before taking over. Taking over is possible even though there are outstanding items, and in this respect it differs from the notion of Practical Completion in JCT forms (12·1).
- Where the Contractor has failed to have the Works ready for taking over by the agreed dates, this fact is certified by the Architect (11·2). Damages may then become payable (11·3 Alternative 1 or Alternative 2).
- Outstanding work at taking over, repairs, replacements and defects may be carried out as instructed by the Architect either during or immediately after the Maintenance Period (12·2).
- The Contractor may be requested to allow the Employer to take over any part of the Works or any section prior to the issue of a Taking Over Certificate (13·1).

3 Control

- Assignment of rights and obligations by either party requires the written consent of the other, but money is expressly excluded from this (9·1).
- The sub-contracting of work by the Contractor requires the written consent of the Architect (9·2).
- Sub-contractors may be named in contract documents. This may be by way of a single name, a list, or a provisional sum. If this arises out of an Architect's instruction, the Contractor is given a right of reasonable objection (9·3, 9·4 and 9·5). The Contractor is responsible for all sub-contractors, including those named (9·9). This extends to all matters of design, compliance with performance specification, and design coordination (9·8).
- The Contractor is clearly made responsible for all management of the Works , and this includes inspection, supervision, planning and superintendence (5·1). The Contractor must appoint as full-time site manager a person approved by the Architect. This person's duties are clearly listed (5·2 and 5·3).
- Facilities are to be provided for access and visits by the Architect, both to the Works and workshops (4·1 and 4·2).
- Instructions and notices given by the Architect must be in writing (23·1) although the contract does allow for oral instructions in an emergency, provided these are confirmed in writing (8·3). Matters on which the Architect is empowered to issue instructions are conveniently listed, and include removal of work or materials from the

site, dismissal from the Works, opening up and testing, variations, and 'any matter connected with the Works'. Immediate compliance by the Contractor is generally required (8·1).

4 Money

- The Contract Sum may be adjusted in accordance with the terms of the contract, and will become the 'Final Contract Sum' – both are exclusive of VAT (clause B of the Agreement, and 15·1).
- A quantity surveyor, if appointed, is named in the Conditions, and his or her duties may be defined (15·2 and 15·3).
- The contract may include for fluctuations based on the ACA index. Deleting this option will make the contract literally 'fixed price' (18·1, 18·2 and 18·3).
- Provisional sums may be included (16·6) although these appear to be only for work or materials in connection with sub-contractors or suppliers.
- Valuations of Architect's instructions will take into account loss and/or expense. Estimates are to be provided by the Contractor before compliance.

  If agreement is not reached, compliance can still be instructed and valuation made based on the Schedule of Rates (if appropriate) or otherwise on a fair and reasonable basis (17·1, 17·2, 17·3 and 17·5).
- Interim payments to the Contractor will be either on the basis of monthly valuations (Alternative A) or by stage payments (Alternative B) (16·1).
- Under both Alternative A and B the Contractor is to present the Architect with an application for payment stating the total due, and supported by documentary evidence as applicable (16·1).
- The Architect is to issue an Interim Certificate within 10 working days of receipt of the application, stating the amount due and the basis of calculation (16·2).
- If the Employer proposes to withhold or deduct any sum, he must give the Contractor written notice showing the amount and grounds for such action not later than five working days before the final date for payment (16·5).
- The Contractor is otherwise entitled to payment of the amount certified. The final date for payment is 10 days from the date of the Interim Certificate (16·3).
- Retention of 5 per cent will normally be retained by the Employer, and in the case of private employers will be placed in a separate bank account without obligation to the Contractor for interest gained (16·4).
- The Contractor has 60 working days following the expiry of the Maintenance Period within which to submit a final account (19·1). The Architect is required to issue the

ACA Form

Final Certificate within 60 days after the Contractor has discharged all his obligations under the Agreement (19·2).

- The Final Certificate must show the amount due and the basis on which the calculation is made. The final date for payment is 10 working days following the issue of the Final Certificate (19·3).

5 Statutory obligations

- Unless instructed to the contrary, the Contractor is required to comply with statutory requirements and to make applications, give notices and pay fees (1·7).
- If there is any apparent conflict between the contract documents and statutory requirements, the Contractor must notify the Architect, who will issue an instruction (1·6). Where the Contractor is responsible for supplying drawings, he also assumes responsibility for ensuring compliance with statutory requirements (2·5).
- The Contractor, where also appointed Principal Contractor, has particular contractual obligations in addition to statutory duties relating to health and safety (26). These include cooperating with the Planning Supervisor, and in respect of the Health and Safety Plan, and the Health and Safety File.

6 Insurance

- In respect of personal injury or death, and damage to property other than the Works, the Contractor gives the Employer an indemnity. This will be reduced proportionately to the extent that the Employer has contributed to the injury or damage (6·3).
- The Contractor will take out insurance cover against these indemnified risks to no less than the sum included in the contract, although of course this will not limit his liability (6·3).
- Insurance of the Works may be by the Contractor (Alternative 1) or by the Employer (Alternative 2). Insurance is to be in joint names. The Conditions do not appear to state specifically what risks are to be covered (6·4).
- Additional insurance may be a requirement for an agreed sum against damage to property (other than the Works) where there is no negligence etc. by the Contractor, e.g. legal nuisance (6·5).
- There is provision for the Contractor to take out design indemnity insurance (6·6).

7 Termination

- The contract includes for termination by the Employer for stated reasons of default. The last one is simply where the Contractor 'shall otherwise be in breach' of the agreement (20·1). The Employer is to serve a default notice, which may be followed by a termination notice. A dispute over this may be referred to adjudication (20·1).

ACA Form

- The contract also includes for termination by the Contractor for stated reasons of default by the Employer. A default notice may be issued to be followed by the termination notice. Adjudication may be used where there is a dispute (20·2).
- Either party is given the option of terminating the Contractor's employment on grounds of insolvency. It is not automatic determination (20·3).
- Either party is given the option of terminating the Contractor's employment for neutral causes (21).
- The consequences of termination, and action to be taken (e.g. payment, possession of site, sub-contracts) is dealt with in detail (22).

8 Miscellaneous

- There is an optional provision for dealing with adverse ground conditions or obstructions on the site (2·6).
- Clause J of the Agreement states that third party rights under the Contracts (Rights of Third Parties Act) 1999 will not apply, but a marginal note states that clause J may be deleted and a clause substituted which lists parties who may be given rights under the contract.
- Confidentiality of documents is assumed (3·3).
- Any drawings provided by the Contractor are his responsibility entirely irrespective of comments or advice by the Architect (3·4).
- Statutory undertakers are to be given access to site (10·3).
- Antiquities etc. are deemed to be the property of the Employer (14·1). The Contractor must report any finds to the Architect immediately (14·2).

9 Disputes

- The Agreement allows for disputes to be resolved by four methods. There is first conciliation (25A), then adjudication (25B), and finally provision for either litigation (Alternative 1) or arbitration (Alternative 2) (25C).
- There are alternative provisions relating to litigation and arbitration depending on whether the proper law of the contract is English law or Scots law (25·11).
- Conciliation can mean reference to a person named in the Agreement, or as otherwise agreed by the parties. Where the parties sign any written agreement on the terms of a settlement, then this is to be regarded as final and binding (25·3).
- Under the Housing Grants, Construction and Regeneration Act 1996 (Part II) the parties have the right to refer any difference or dispute to adjudication. The adjudicator may be named in the Agreement, or be otherwise appointed in accordance with the CIC Model Adjudication Procedure (25·5).

ACA Form

- Adjudication is to be conducted according to the CIC Model Adjudication Procedure, and the adjudicator's decision will be final and binding at least until the dispute is finally determined by arbitration or in the courts (25·7).
- Arbitration is subject to the provisions of the Arbitration Act 1996 if under English law, or its counterpart if under Scots law. If the parties do not agree over the name of the arbitrator, the appointor is to be the President of the Chartered Institute of Arbitrators (25·9).

ACA Form

This contract?

If considering using the ACA form remember that:

It is intended for lump sum contracts, and can be used regardless of sector although certain clauses may need to be deleted if the Employer is a local authority. It may be used with or without a bill of quantities. Work needs to be fully designed and reasonably well documented at tender stage, although there is alternative provision for further detailing by either the Architect or the Contractor. Possession and completion may be in respect of a single contract period or for sectional completion in accordance with a Time Schedule. The Employer is required to appoint an Architect or Supervising Officer, and may appoint a quantity surveyor.

The form is stated to be suitable under Scots law, and there are alternative clauses relating to litigation and arbitration. There is no reference to its suitability under the law of Northern Ireland.

A notable feature of this form is the range of alternative clauses available to cover most situations, for example relating to responsibility for further necessary information including drawings; priority of contract documents; adverse ground conditions; insurance of the Works; liquidated damages; grounds for extensions of time; payment periodic or by stages; dispute resolution; possession and completion by sections. The form appears to deal in a very simple way with naming of sub-contractors, design by sub-contractors, and the procedures are kept to a very minimum. This flexibility can be achieved all within the one document, and without the need for supplements. The only bolt-on document is the ACA Conciliation Procedure 1998.

Completing the form requires particular care over selecting the appropriate combination of alternatives, and in completing the Schedules. The form has a logical structure and the language used is straightforward English. However, some of the terms used may be unfamiliar to regular JCT users.

The Contract Administrator is given considerable authority. The procedural rules are straightforward. ACA publishes standard forms and certificates for use with the ACA Form.

This is a different form for traditional procurement, attractive, concise and modern. It is probably best suited to contracts which are middle range in value, and in this respect could be compared with JCT IFC98. It seems not to have attracted adverse reaction from contractors despite its unilateral production.

7 Traditional procurement: shorter lump sum forms

ACA Form

Related matters

Documents

ACA Form of Building Agreement 1982; Third Edition 1998 (2000 Revision)
ACA Form of Sub-Contract 1982; Third Edition 1998 (2000 Revision)
ACA98 Appointment of a Consultant Architect

Notes

Guide to the ACA Form of Building Agreement.

GC/Works/2

The Stationery Office

GC/Works/2 (1998)

Background

This is sub-titled as a Contract for Building and Civil Engineering Minor Works, but it is more in the nature of an intermediate form standing between the major works GC/Works/1 form and the GC/Works/4 small works contract. It is intended for contracts of between £25,000 and £200,000 in value, and for demolition works of any value. It replaces the former GC/Works/2 (1990) and the old small works C1010 (1990).

It is published as a two volume package, General Conditions, and Model Forms and Commentary. The former includes a disk (Word Perfect 6.1) of the 18 Model Forms.

The form is for use with lump sum tenders invited on the basis of Specification and drawings only – without provision for bills of quantities, and with the optional requirement for the Contractor to submit a Schedule of Rates to enable fair valuation of any variations.

Nature

The General Conditions run to some 45 pages and follow the well established GC/Works pattern of clear graphic style, straightforward language and standard terminology. The Introduction and contents list are followed by the Conditions of which there are 48, structured under nine headings.

In common with other GC/Works forms, there is a very useful Schedule of Time Limits and an alphabetical index. Use of the Model Forms is essential. Model Form 1 contains the Abstract of Particulars and Addendum; Model Form 2 is an Invitation to Tender and Schedule of Drawings; Model 3 is a Tender and Tender Price Form; Model Form 4 carries details of Insurance Documents; Model Form 5 is a Performance Bond; Model Form 6 is a Parent Company Contract Performance Guarantee; Model Form 7 relates to the Appointment of an Adjudicator; and Model Form 8 is an 'Order to Proceed'. The latter is in effect an instruction to the Contractor to proceed with the Works on a specified date. There is no separate Contract Agreement.

Use

GC/Works/2 has relatively limited provisions, but almost certainly adequate for the kind of operations intended. As stated in the Commentary it is necessary when considering a choice of this form compared with, say, GC/Works/1, to establish whether certain features (e.g. sectional completion) are contractual requirements. If so, then this less comprehensive document might not prove suitable. The choice must be determined by the circumstances, the nature of the project, and the balance of risks to be covered.

GC/Works/2

As there is no formal Contract Agreement, a contract will be brought into existence when the Employer signifies acceptance of the Contractor's tender. This should be in writing to bring certainty to the arrangement, particularly as the result will almost invariably be a 'construction contract' as defined in the Housing Grants, Construction and Regeneration Act 1996 (Part II). GC/Works/2 makes no provision for a quantity surveyor, but a Project Manager and Planning Supervisor should be named in the Abstract of Particulars, and the name of an adjudicator and an arbitrator may also be entered.

Synopsis

1 Intentions

- A fair dealings provision is included (1A).
- The Contractor is to carry out the Works in accordance with the Contract Documents (2), and instructions of the Project Manager (25). The Works are as described in the Specification and shown on drawings, and will include all modified or additional works to be executed under the contract (1). They are to be executed in a workmanlike manner and to the satisfaction of the Project Manager (19).
- The Contractor is deemed to have satisfied himself about all matters which might affect carrying out the Works. No additional payment will be allowed because of misunderstanding (4).
- All 'Things' selected by the Contractor for incorporation in the Works, will be as described in the Specification and drawings and must conform to the requirements of the contract (19[2]).
- The 'Contract' means the written agreement concluded by tender, acceptance, Conditions, Abstract of Particulars, Specification and drawings, etc., all taken together (1). In case of discrepancy, there is a detailed hierarchy (2).
- Further drawings, details, instructions etc. may be issued by the Project Manager from time to time during the Works (25).

2 Time

- Possession is given to the Contractor by written order to proceed. He must thereupon commence on a certain date, and 'proceed with diligence' and complete by the Date for Completion (21[1]).
- A reasonable extension may be awarded by the Project Manager, only for circumstances wholly beyond the Contractor's control (23).
- The Works must be cleared of rubbish and delivered up to the Project Manager's satisfaction by the Date for Completion (21[4]).

GC/Works/2

- Defects which appear in the Maintenance Period stated in the Abstract of Particulars must be made good to the satisfaction of the Employer at the Contractor's expense (9).

3 Control

- The Contractor cannot assign the contract or any interest without written consent of the Project Manager (44). Sub-contracting requires the written consent of the Project Manager.
- There is no provision for sub-contractors and suppliers to be nominated by the Employer or Project Manager.
- The Project Manager may issue instructions on any matter necessary, including variations (25). Notices must be in writing, and oral instructions confirmed in writing. The Contractor must comply forthwith.
- Instructions which result in a variation will be valued wherever practicable by prior quotation and agreement, or a relevant charge for daywork, or by the Project Manager, on fair rates and prices, and will include for disruption or prolongation costs (26).
- There is no reference to setting out, nor to the Contractor's person-in-charge.
- The Project Manager and his representative are empowered to order tests, and the cost will be borne by the Contractor where things are not in accordance with the contract (19).
- The Employer may carry out work directly during the time the Contractor is in possession (46).

4 Money

- The Contract Sum is defined (1), and the 'Final Sum' will be adjusted in accordance with the contract.
- The Project Manager must give written instructions before provisional sum work is commenced (45). It will be valued as provided for in the contract (26).
- The Contractor, following certification by the Project Manager, shall be entitled to monthly payment of advances on account (30). Valuations will be prepared by the Contractor and, if agreed by the Project Manager, the sum due will be certified by him.
- The Contractor may have to allow credit for old materials. 97 per cent of the value of work executed and value of Things for incorporation will be paid during progress of the Works (30).
- On completion, the Contractor can expect to be paid what the Employer estimates the Final Sum to be, less half the retention (31[1]).

GC/Works/2

- The Project Manager will send the Contractor a draft final account within six months after completion, which is certified by the Project Manager. The Final Sum will be paid when the Project Manager certifies that the Works are in a satisfactory state following the Maintenance Period (31).
- The contract fully complies with the Housing Grants, Construction and Regeneration Act 1996 provisions concerning payment procedures.

5 Statutory obligations

- The Contractor is required to give all notices, pay any fees and charges, and make and supply all drawings required to support such notices, arising from statutory obligations (6). This includes the CDM Regulations.
- The Contractor is required to comply with any statutory regulations relating to storage and use of all things brought on the site (7[2]).
- The Contractor is required to comply with applicable occupier's rules and regulations in respect of the site (10).

6 Insurance

- The Contractor is required to take out and maintain employer's liability insurance in respect of his employees on site, and is advised to insert a similar provision in sub-contracts (5[1]).
- The Contractor shall insure in joint names against all risks damage to the Works and Things for which the Contractor is responsible. This will be for full reinstatement value plus professional fees (5[2]).
- The Contractor indemnifies the Employer in respect of injury to persons, loss or damage to third party property, loss of profits or use, and will cover this by insurance (5[2] and 8).
- The Employer is to bear the certain specified risks arising from work to existing structures, and may elect to take out insurance (5[6] and 8).

7 Termination

- The Employer has the right to determine the contract for specified grounds including insolvency of the Contractor (38).
- Matters following determination (payment, completion, removal, etc.) are dealt with in Condition 39. The Project Manager must certify the cost of completion.
- The Contractor may determine the contract for specific grounds, including insolvency of the Employer (40).
- Either party may determine the contract following suspension of the whole or substantially the whole of the Works (41).

GC/Works/2

8 Miscellaneous

- A full list of definitions is included (1).
- There is no express reference to rights of third parties, or to contracting out of the Contracts (Rights of Third Parties) Act 1999.
- The Contractor has obligations concerning watching, lighting and protection (7) and for removal of rubbish (21[4]).
- Security matters such as admission to the site (14), taking photographs (16), passes (15), and Official Secrets (17) may give rise to relevant obligations.

9 Disputes

- There is provision for adjudication for the resolution of any dispute arising out of the contract (42). There are precise requirements for the notice of referral and the procedures to be followed. The adjudicator may be named in the Abstract of Particulars, and his decision is binding on the parties, at least until finally determined by arbitration.
- Arbitration is given as the forum for final resolution of disputes and the arbitrator is given wide powers under the contract (43).

GC/Works/2

This contract?

If considering using GC/Works/2 remember that:

It is intended for use when lump sum tenders are being invited on the basis of drawings and a Specification only. Its use is no longer confined to central government departments. The Employer is required to appoint a Project Manager who is given considerable powers to act for the Employer, subject to excluded maters which may be listed in the Abstract of Particulars. There is no requirement for a quantity surveyor, and functions normally ascribed to him are the responsibility of the Project Manager.

The form may be used in England and Wales, or under the law of Northern Ireland, or under Scots law. The arbitration provisions appropriate to the latter are set out in Condition 43.

The Conditions are general, and the Abstract of Particulars allows for some flexibility in tailoring the contract to the nature of the intended work. There is no provision for sectional completion. There is no provision for nomination of sub-contractors or suppliers by the Employer. The two volume presentation is comprehensive, and the Model Forms should be used as some of them carry information which is supplemental to the Conditions. The Commentary is practical and particularly informative.

Contract administration should be straightforward, always provided that the administrator takes the trouble to become thoroughly conversant with the terminology, the procedures, and the time limits. Rather unusually but very sensibly, progress meetings become contractual obligations, although there is no requirement for a programme.

This is an attractive form which has many features not found in other short forms. The result is a fairly substantial document which can cover a wide range of work which is too complex for GC/Works/4, but not justifying use of GC/Works/1. It may also be seen as an alternative to using JCT forms MW98 and IFC98.

Related matters

Documents

GC/Works/2 (1998) Contract for Building and Civil Engineering Minor Works in two volumes: General Conditions; Model Forms and Commentary

GC/Works/4

The Stationery Office

Background

This is sub-titled as a Contract for Building, Civil Engineering, Mechanical and Electrical Small Works, and obviously intended as a general purpose document to cover small works of a varied nature. It is intended for contracts of up to £75,000 in value, and replaces Form C1001 (1990).

It is published as a two volume package; General Conditions and Model Forms and Commentary The former includes a disk (Word Perfect 6.1) of the 13 Model Forms.

The form is for use with lump sum tenders invited on the basis of Specification and drawings only, without provision for bills of quantities.

Nature

The General Conditions run to some 18 pages and are set out in clear graphic style, straightforward language and GC/Works terminology. There is a contents list followed by the Conditions of which there are 31 structured under seven headings.

As with other GC/Works forms, there is a Schedule of Time Limits. Use of the Model Forms is essential. Model Form 1 is the Abstract of Particulars; Model Form 2 is an Invitation to Tender and Schedule of Drawings; Model Form 3 is Tender and Tender Price Form; Model Form 4 is the Adjudicator's Appointment; Model Form 5 is the Order to Proceed; Model Forms 6 to 13 are contract administration forms.

Use

GC/Works/4 has provisions which should be sufficient for most small contracts where services installations also form part, but choice should ultimately depend on the nature of the intended works. The form accepts that either all the CDM Regulations will apply, or that only CDM Regulations 7 and 13 will be applicable. It also takes account of the Housing Grants, Construction and Regeneration Act 1996 (Part II).

Brief synopsis of Conditions

- A fair dealings provision is included (1A).
- Conditions of contract prevail in the event of discrepancy, except where special supplementary conditions are included in the Abstract of Particulars (2·1).
- Specification takes precedence over drawings unless otherwise instructed (2·2).
- Contractor responsible for care of site and Works, including sole responsibility for protection, security, lighting, and watching over the site and Works (5).

GC/Works/4

- Contractor indemnifies Employer in respect of any loss or damage which arises out of work under the contract (6). There is no provision for insurance in this contract.
- Making good defects in the Maintenance Period will only apply if stated in the Abstract of Particulars (7).
- The usual GC/Works provisions for matters such as occupier's rules, site admittance, passes, photography and Official Secrets, are included in a truncated form.
- The Contractor's dates for commencement and completion will be stated in the Abstract of Particulars and subject to the order to proceed (15).
- The Works must be cleared of rubbish and delivered up to the Project Manager's satisfaction on completion (15).
- A reasonable extension of time may be awarded by the Project Manager, but only for circumstances beyond the Contractor's control.
- The Contractor cannot assign or sub-let without the written consent of the Project Manager (30).
- There is no provision for provisional sums.
- The Project Manager may issue instructions on any matters necessary, including variations (17).
- Valuation of variation instructions whenever practicable will be by prior quotation and agreement, or by the Project Manager on the basis of fair rates and prices (18).
- There are no contract provisions relating to progress meetings.
- There is no provision for prolongation and disruption.
- The Employer may carry out work direct during the time the Contractor is carrying out the works (31).
- The contract price (tender as accepted) may be adjusted in the final account (21).
- Payment to the Contractor will be after completion, unless the Abstract of Particulars states that there will be advances on account (20). This will mean one-third payable after one-third of the work completed, with the remainder after completion. Applications must be made by the Contractor to the Project Manager, and are subject to certification by the Project Manager (22).
- In the event of failure to complete to time, the Contractor may be liable for liquidated damages, or damages at large as stated in the Abstract of Particulars (26).
- The Employer has the right to determine the contract for specified grounds, including insolvency of the Contractor (27).
- The Contractor has no rights to determine the contract.

GC/Works/4

- Adjudication is the only provision for resolving disputes (28). The decision of the adjudicator is binding until finally determined by arbitration, but the form does not include for arbitration. Alternatively the parties may agree to accept the decision of the adjudicator as final determination.
- There is no express reference to rights of third parties or to contracting out of the Contracts (Rights of Third Parties) Act 1999.

This contract?

If considering using GC/Works/4 remember that:

It is intended for use when lump sum tenders are being invited on the basis of drawings and a Specification only. Its use is no longer confined to government departments, although some provisions will obviously not be relevant to the private sector client (for example the recovery of sums by the Crown under Condition 24). The Employer is required to appoint a Project Manager, but there is no stated function for a quantity surveyor.

The form may be used under the law of England and Wales, or under the law of Northern Ireland or under Scots law. The adjudication provisions appropriate to the latter are covered by Condition 29.

The Conditions are relatively brief and although limited, are adequate for most small works. The Abstract of Particulars becomes an important document. The two volume pack is attractively presented although perhaps a trifle over-sophisticated for this class of work. The Model Forms should be used.

Contract administration should be straightforward, and the procedures are simple. The terminology is typical GC/Works but clear.

An interesting basic alternative small Works Contract, particularly where services installations form part of the works. Likely to appeal most to those already familiar with other GC/Works contracts.

Related matters

Documents

GC/Works/4 (1998) Contract for Building and Civil Engineering, Mechanical and Electrical Small Works in two volumes: General Conditions; Model Forms and Commentary

NEC

The Institution of Civil Engineers
New Engineering Contract Document

Engineering and Construction Short Contract First Edition (July 1999)

Background

Although the New Engineering Contract of 1993 was at first promoted as a contract which could be used across the whole spectrum of projects, the Latham Report stated the need for 'a simpler and shorter minor works document'. Such a document, the Engineering and Construction Short Contract was published in July 1999 as one of the NEC family.

Nature

The form is obviously a derivative of the ECC contract, but it has been very thoughtfully structured. It is much more than simply a paired down version of the major document. The important material peculiar to a particular contract is placed right up-front (the converse of the major document) and comprises:

- Contract Data: dates; damages; interest on late payments; limit of Contractor's liability; insurance obligations; name of adjudicator and arbitration procedures (if relevant). Additional Conditions may be incorporated and if so they are to be listed.
- Contractor's Offer: which will indicate whether this is a lump sum or a remeasurement contract.
- Employer's Acceptance: signed and dated. The offer and acceptance bring about the contract of course.
- Price List: a schedule of the quantity, rates and prices of items.
- Works Information: description of the Works which the Contractor is to carry out, and any work which the Contractor is to design; list of drawings; list of specifications applicable; constraints on the Contractor such as sequence, timing, methods, and conduct of work; requirements for a Programme, such as form, content, submission and updating; and services to be provided by the Employer.
- Site Information: ground conditions, access, position of adjacent structures etc.

The Conditions of contract, which occupy only 11 pages, are set out in a similar although not identical way to those in the major form. The clauses are of necessity briefer than those in the major contract, and some provisions are omitted entirely, such as testing, health and safety compliance etc. There is no express provision for a

Contract Administrator or Project Manager, but the Employer may delegate the authority for taking empowered actions.

The language used in the form has attracted a 'Clear English Standard' award, the terminology is consistent with other documents in the NEC family, and the balance is generally attractive. The only provisions which stand out as being particularly demanding in terms of management expertise are those which concern the Compensation Events, which are almost as complex as those in the major form.

The contract as printed does not appear to take account of the Housing Grants, Construction and Regeneration Act 1996. At the end of the Conditions, there are replacement clauses concerning adjudication, but if the particular contract is a construction contract to which Part II of the 1996 Act applies, then the payment provisions would also need attention, unless the Scheme for Construction Contracts is left to apply. The clauses in Addendum Y(UK)2 for the Engineering and Construction Contract could be adapted and incorporated by reference in the Contract Data. The Addendum Y(UK)3 which takes account of the Contracts (Rights of Third Parties) Act 1999 may also be incorporated, and this Addendum is stated to be applicable to the short contract.

Use

This short contract has been specifically produced for use with straightforward work, for which sophisticated procedures are not necessary, and where the risks are relatively low. There is no restriction on the value of contract work for which it might be suitable, and it is the nature of the intended work which must be the determining factor.

Brief synopsis of Conditions

1 Intentions

- The parties act in a contractual spirit of mutual trust and cooperation (10·1).
- The terms used in the contract are helpfully and fully defined (11).
- Communications are to be in writing, and if a period for reply is stated in the Contract Data, then within that period (12).
- The Employer may give instructions which change the Works Information (13·2).
- The Contractor must obey empowered instructions issued by the Employer (13·1).
- The Employer must allow the Contractor access and use of the site (note: not necessarily possession), and provide the services stated in the Works Information (14).
- Early warning is a significant requirement in the contract (15).

NEC

2 Time

- Starting date and completion date are entered in the Contract Data. The Contractor may not start before, and completion may be on or before the completion date (30·1).
- The Contractor must submit a forecast of the date for completion to the Employer each week (30·2).
- The Employer may instruct the Contractor to stop or restart work (30·4).
- The Contractor must submit a Programme as required in the Works Information (31).
- Where delay occurs due to certain intervening events, these may constitute Compensation Events (60). There are 13 listed, mostly arising from action or inaction by the Employer, but also including some neutral causes. These are for weather (measured in ratio of days lost against contract period); physical conditions beyond those which could reasonably have been expected; and an event which delays completion by more than two weeks and which could not normally be expected to occur. This would presumably not allow for strikes or lock-outs.
- The procedures for notifying Compensation Events, quotations for changes to prices or rates and the mechanism for assessing the events are almost as full for this short contract as those in the major form (62 and 63).

3 Control

- The early warning obligation can constitute a control mechanism (15).
- The Contractor cannot start work on which he has designed, before the Employer has accepted that the design complies with the Works Information (20).
- The Employer may delegate any actions, and may cancel any delegation (13·4).
- The Employer may instruct the removal of an employee (21·3).
- The Contractor must provide access for the Employer and others to work and to stored plant and materials (22).
- There is no consent required for sub-contracting, but the Contractor remains wholly responsible (21).
- There is no provision for testing.
- The Employer may instruct the Contractor to search for defects (40·1) but whether or not the Employer notifies the Contractor of a defect, he is obliged to correct them (41).
- After completion the Contractor is to correct notified defects before the end of the Defects Correction Period (stated in weeks in the Contract Data) (41·2), and the Employer is to issue a Defects Certificate (41·3).

NEC

- Uncorrected defects permit the Employer to engage others to carry out remedial work, the cost of which can be charged to the Contractor (42).
- Until the Defects Certificate has been issued, the Contractor is liable for replacement in the event of loss of plant or materials, and damage to the Works (43·1).

4 Money

- The Contractor is to make application for payment by the assessment day (that is the date stated in the Contract Data) once a month (50·1).
- The Contractor's application will include details of how the amount has been assessed and is for the amount due (50·2).
- Retention is held according to the percentage stated in the Contract Data, half is released on completion and the remainder after the Defects Certificate is issued (50·6).
- The Employer is to pay within three weeks after the next assessment day which follows receipt of application for payment from the Contractor. Interest becomes payable on late payments (51).

5 Statutory obligations

- There are no express obligations relating to statute or regulations, but these would be implied. In the event of a breach of statutory duty by the Employer, the Contractor is indemnified against claims or proceedings and costs arising (81·1).
- In particular there is no express reference to health and safety and the CDM Regulations, other than providing a reason for termination by the Employer if the Contractor substantially breaks a regulation (90·3).

6 Insurance

- The Contractor is to provide in joint names the insurances stated in the Insurance Table (82·1).
- The Employer may provide insurance where this is stated in the Contract Data, and where this occurs it is not provided by the Contractor (82·1).
- The insurance is for replacement cost in respect of loss or damage to plant, materials and the Works, and extends from starting date to Certificate of Completion in the case of plant and materials, and Defects Certificate in the case of the Works.
- Insurance may also be required in respect of the Contractor's liability for damage to property other than the Works, and injury or death to persons for the minimum cover as stated in the Contract Data.
- The Contractor's liability for loss or damage to the Employer's property is limited to the amount stated in the Contract Data (80·1).

NEC

7 Termination

- Both the Employer and the Contractor have the right to determine (and this presumably means the contract, although not expressly stated).
- There are eight reasons stated which allow for termination, including insolvency of the other party (90·2). The Employer may also terminate 'for any other reason' (90·6).
- The Employer must issue a Termination Certificate, after which the Contractor must cease work (90·1).
- On termination the Contractor must leave the site (91·1) and the Employer may have the Works completed by others.
- Amounts due on termination are as stated, according to the reasons which apply (92).

8 Miscellaneous

- Use of the NEC Engineering and Construction Short Sub-contract would seem to be necessary.
- The Addendum Y(UK)2 is not applicable for use with the short contract, but payment provisions no doubt could be adapted and incorporated by reference in the Contract Data.
- The Addendum Y(UK)3 on third party rights is stated as being suitable for use with the short form.
- Clauses 93 to 95 on dispute resolution in the short form should be replaced by clauses 93UK to 95UK, for use in the United Kingdom where the Housing Grants, Construction and Regeneration Act 1996 applies.

9 Disputes

- Settlement of disputes is to be by reference to adjudication (93).
- The adjudicator may be named in the Contract Data.
- The adjudicator's decision is final and binding unless and until referred to a further 'tribunal'. The Contract Data should state whether arbitration is the tribunal, and if so, what the arbitration procedure is to be.

NEC

This contract?

If considering using the ECC short contract remember that:

It is intended for use as a lump sum or remeasurement contract for minor building and engineering works. Care is needed at the outset to establish that the Conditions, which are short but generally commensurate with work of this nature, will be adequate for the intended works. The agreement is based on the Offer and Acceptance near the front of the contract form. Contract Data and Works Information are particularly important and require full entries and careful checking by both parties.

The law of the contract is that applying to where the site is. There is no express reference to the law of Northern Ireland or to Scots law. Presumably any consequential amendments necessary to the Conditions would have to be included in Contract Data.

The document is rather deceptive in its apparent simplicity because whilst some of the Conditions are admirably brief and straightforward, others seem to be lightweight by comparison with equivalent clauses in other minor works forms, for example over sub-contracts or insurance. The clauses for Compensation Events, however, seem rather out of proportion for a short form, although it must be difficult to abbreviate this NEC concept of dealing with additional time and money.

The use of the present tense, as with other NEC documents, is somewhat difficult to cope with at first, and tends to give an air of imprecision. The terminology is standard NEC, and the form has a full definitions clause. Recent UK legislation, where applicable to small-scale works, seems not to have been taken into account. This keeps the form short, but incorporation by the use of Addenda can be a trifle confusing.

Most employers would probably welcome professional advice when it comes to completing the form, and certainly the Compensation Events procedures suggest the need for an employer's 'delegate' for contract administration.

However, for committed and experienced NEC users, this form must be seen as a welcome addition to the family.

Related matters

Documents

Engineering and Construction Short Contract First Edition (July 1999)
Addendum Y(UK)3 (April 2000)
NEC Partnering Option, Option X12 (June 2001)
Engineering and Construction Short Sub-Contract

Notes

Engineering and Construction Short Contract Guidance Notes and Flow Charts

JCLI Agreement

The Joint Council for Landscape Industries (JCLI)

JCLI Agreement for Landscape Works

Background

The JCLI originally had two standard forms for landscape works based on JCT63 With and Without Quantities. In 1978 these were replaced by a single form modelled on the then current JCT Minor Works Agreement 1968. Their new form adopted a section headed format (two years ahead of the JCT) and incorporated additional clauses of specific relevance to landscape works. These included such matters as the failure of plants, malicious damage and theft, and additional retention to cover plants dead at Practical Completion. The form also contained provisions for partial possession, Prime Cost sums, objections to nomination, delays by the Employer and full fluctuations, none of which were included in the JCT Minor Works form. This form was considerably revised in 1987, again in 1989, and is currently in a 2002 revision of a 1998 edition. Landscape contracts will mostly be 'construction contracts' and need to take account of the Housing Grants, Construction and Regeneration Act 1996 provisions.

In 1985 the JCLI also published a Supplement to the then JCT Intermediate Form (IFC84) for use on larger landscape contracts. This was thought desirable because at the time some local authorities were reluctant to use the standard JCLI Agreement on contracts in excess of £75,000.

JCLI also publishes a JCLI Agreement for Landscape Maintenance Works for use where the landscape contractor is to be responsible for the care of trees, shrubs and grass after Practical Completion.

Brief synopsis of Conditions

- As with all 'construction contracts' adjudication is a statutory right, and payment procedures must comply with the provisions of the Housing Grants, Construction and Regeneration Act 1996.
- The form refers to 'the Contract Administrator' and sets out respective duties (1·2).
- There is provision for contract bills, and these are to be prepared in accordance with SMM7 unless stated otherwise (1·8).
- The applicable law, whatever the nationality of the Employer, and wherever the Works are situated, is the law of England (1·7).
- Works may be subject to the 'CDM Regulations' (Fifth Recital). If mainly soft landscape works then this is likely not to be 'construction work' and Alternative A will apply. If landscape work includes earthworks, hard landscaping, drainage, demolition etc. this might constitute 'construction work' in which case Alternative B will apply.

JCLI Agreement

- There is provision for partial possession (2·6) but no stated procedures.
- There is provision for replacement of plants etc. which fail before Practical Completion (2·7).
- There is provision for the maintenance of trees, shrubs, plants, grass, etc. post-Practical Completion (2·8) either by the Contractor (A) or the Employer (B).
- There is provision to cover making good, following malicious damage or theft prior to Practical Completion, either entirely at the Contractor's own risk (2·9A), or covered by a provisional sum (2·9B).
- Although the form does not refer to any procedures or rules for nominating sub-contractors, there is reference to instructions by the Contract Administrator on the expenditure of Prime Cost sums and a right of reasonable objection to a nomination is given to the Contractor (3·8).
- The contract may be fixed price (4·7), or subject to tax etc. changes (4·6).
- Adjudication is provided for (8·1), and disputes may be finally determined by arbitration or by litigation (8·2 and 8·3).

JCLI Agreement

This contract?

If considering using the JCLI Agreement remember that:

It is intended for new landscape work, both soft and hard landscaping. It covers work up to Practical Completion, but maintenance is a matter for the JCLI Agreement for Landscape Maintenance Works. It is in one version only, and for use in either private or public sectors. There is provision for bills of quantities. The Employer must appoint a Contract Administrator (not necessarily a landscape architect) and may appoint a quantity surveyor.

Subject to anything to the contrary, the applicable law of the contract, whatever the nationality of the parties, and wherever the Works are situated, is the law of England. For landscape contracts in Scotland, amendments to the dispute resolution provisions would be required, or preferably the SBC Minor Works Contract should be considered.

There is no provision for phased completion, fluctuations, or landscape design by the Contractor. There is provision for partial possession, nomination of sub-contractors, and clauses dealing with the failure of plants, malicious damage or theft, and post-Practical Completion care. The form is drafted to include compliance with the CDM Regulations and the Housing Grants, Construction and Regeneration Act 1996.

When completing the form decisions are required relating to insurance of the Works, damages for non-completion, and post-Practical Completion care. There is no conventional appendix to the form, and entries are to be made in the text of the Conditions.

If acting as Contract Administrator, a non-expert in landscape operations might find difficulties over Practical Completion, defects liability obligations and staggered release of retention. The procedural rules in general are straightforward but if the agreement is used in parallel with building work being carried out on the same site by a Principal Contractor, care is advisable over coordination of contract periods, Defects Liability Periods, health and safety documents, indemnity and insurance.

The JCLI issues loose-leaf revisions and Practice Notes which give useful advice on matters peculiar to landscape work such as plant specification, watering, frost damage, temporary protection, liability for plant failures and multiple Defects Liability Periods.

The wording of the Conditions closely follows those of the JCT form MW98 and therefore the structure, language and terminology will be familiar to many. For large hard landscape work it might be worth considering the use of the JCT form IFC98 suitably modified as an alternative.

JCLI Agreement

Related matters

Documents

JCLI Agreement for Landscape Works 1998 Edition (2002 Revision)
JCLI Agreement for Landscape Maintenance Works 1998 Edition (2002 Revision)

References

JCLI Practice Note No 5 (February 2002)
JCLI Practice Note No 7 (February 2002) for Landscape Maintenance Works

ASI Forms

Architecture and Surveying Institute

ASI Forms of Contract

Background

The Architects and Surveyors Institute (as it was then called) came into being in 1989 as a result of the amalgamation of the Faculty of Architects and Surveyors and the Construction Surveyors Institute. Some of the forms currently produced for members were similar to those which originally appeared under the Faculty impress. However, a new generation of ASI documents has now appeared, and these go a long way to constituting a 'complete family of interlocking documents' as advocated by the Latham Report.

The name of the ASI was changed to Architecture and Surveying Institute in 1998, following a legal ruling which brought stricter control over the use of the term 'Architect'.

ASI forms exist to cover a wide range of building operations as befits an Institute which counts architects, architectural technicians and all branches of the surveying profession among its members. The three main contracts have each been approved by the British Institute of Architectural Technologists. The forms are for use with traditional methods of procurement, taking a lump sum as the contract basis.

Nature

All ASI forms are plainly worded, and in the main are commendably free from legal jargon. They were given a face-lift in 2000. Both contracts and sub-contracts are structured to a common pattern, with the clauses in the main contract Conditions grouped under 11 headings. The contracts currently published are as follows:

ASI Building Contract (2000 Edition): suitable for all types of works, but primarily intended for larger or complex jobs of all types. It may be used either with or without bills of quantities. ASI domestic sub-contract forms and nominated sub-contract forms are available.

ASI Small Works Contract (2000 Edition): intended for use with smaller works e.g. private houses, alterations or extensions, with the work shown and described in drawings and/or Specification or Schedule of Works. Supplementary Conditions permit the use of bills of quantities if required. This contract also makes provision for nominated sub-contractors and fluctuations if required.

ASI Minor Works Contract (2000 Edition): for use with straightforward jobs or minor works of a relatively simple character carried out on a lump sum basis. It is not suitable for use with bills of quantities or where it is intended to nominate sub-contractors.

ASI Forms

ASI Mini Contract: in two versions, for Home Improvement Agencies and General Use (1998 and 2000 Editions respectively).

The forms referred to in this section are the three main contracts listed above. See Chapter 8 below for details of the ASI Mini forms of contract.

Use

ASI forms are suitable for use in the private sector or by local authorities. A Contract Administrator is needed in all cases, but that person may be an architect, surveyor or engineer. The Contract Administrator is referred to in the Conditions as 'the Architect' in the Building Contract and Small Works Contract, and 'the Adviser' in the Minor Works Contract – regardless of profession! (A footnote reminder on p. 1 states that the term 'Architect' must be deleted on that page unless the Contract Administrator is a Registered Architect, but nevertheless appears to accept the term when relating to contract administration thereafter.)

All the forms commence with entries which identify the parties and date of the contract. There are brief Recitals, and the Agreement in which details relevant to the particular contract must be entered. There is provision for attestation to be under hand and not as a deed, or as a deed. Each form has appendices peculiar to the specific contract, and each has an index list of clauses. The forms appear to comply with the provisions of the Housing Grants, Construction and Regeneration Act 1996 in respect of adjudication but not payment, and if so then the relevant provisions of the Scheme for Construction Contracts would therefore apply.

Brief synopsis of forms

ASI Building Contract (2000)

- A headnote indicates that the form is intended for larger or complex jobs, and for use with or without bills of quantities.
- Contract documents are drawings, and bills of quantities or Specification. If the latter then the Contractor must supply a Schedule of Rates.
- Details relating to the particular contract are to be entered in the Agreement.
- The person appointed by the Employer as Contract Administrator is referred to as 'Architect'. If that person is 'Surveyor/Engineer' this will be an entry in the Recitals and the wording of the Conditions is deemed to have been changed.
- The obligations of the Contractor and the role of the Architect are well described (1 and 2).
- The Contractor must submit a Programme/progress chart within three weeks of entering into the contract (4·11).

ASI Forms

- Key contract dates are entered in the Agreement, and the Architect may award at his or her absolute discretion extensions of time for listed admissible causes of delay (4·4).
- The contract requires Practical Completion (leaving 'only acceptable minor items to be completed'), and can also accommodate Part/Sectional Completion as the 'Employer may require or agree with the Contractor' (4·6).
- Failure to meet the completion date will result in payment of liquidated damages, as confirmed in writing by the Architect (4·5).
- The Employer may appoint a Clerk of Works empowered to issue directions, but which must be confirmed as Architect's instructions (2·6).
- Domestic sub-contracting is permitted, subject to approval by the Architect (5·1).
- There is provision for nominated sub-contractors and Prime Cost suppliers, subject to stated conditions (5·3 and 5·4).
- The Contract Sum is VAT-exclusive. Fluctuations in scheduled rates and prices for labour and materials may be included, or there can be adjustment by use of formula rules. Otherwise only changes in statutory charges will be allowable (7·3).
- The Architect is empowered to order variations, and the order will also stipulate the basis for valuation (i.e. either according to the priced document; or rates approved by the Architect; or a quotation approved by the Architect; or as daywork). Note that if there is no quantity surveyor, then measuring is the responsibility of the Contractor (7·2).
- Architect's Interim Certificates are normally issued at monthly intervals, with payment by the Employer due within 14 days (6·3). Amounts payable to nominated sub-contractors must be identified in the Certificates, and the Contractor must provide proof of payment. Failure to comply means direct payment by the Employer (6·81).
- Architect's Final Certificate is to be issued 'as soon as practicable' after any defects have been made good following the end of the Defects Liability Period – normally six months (6·9).
- No certificate is to be taken as conclusive evidence that work, materials or goods are in accordance with the contract (6·91).
- The Contractor must comply with all legal and statutory requirements, including giving necessary notices and paying fees and charges legally demandable (1·3). If changes from contract documents become necessary for reasons of statutory compliance, the Architect must receive written notice from the Contractor (1·31).
- Compliance with health and safety CDM Regulations becomes a contractual obligation as well as a statutory one (Agreement A1 and 1·3).
- Insurance of new work in joint names, against risk of damage by listed perils, is by the Contractor (8·1), although the Employer may elect to carry this insurance. Insurance

of existing buildings against risk of damage by listed perils is at the sole risk of the Employer (8·2).

- There is provision for determining the employment of the Contractor for listed reasons of default on the part of either Employer or Contractor. The procedures are clearly set out, and if determination occurs then actions required in respect of removal from the site, completion of the Works, and settlement of claims, are all described (9·1 and 9·2).
- Disputes or differences are to be submitted to the Architect initially for his or her decision (10·1). If not resolved then the dispute may proceed to arbitration (10·2). The contract lists the several methods and rules which can apply, including a shortened arbitration procedure (10·3). There is also a section on the right of parties to refer a dispute to adjudication and the procedures relating to the appointment of an adjudicator and conduct of the adjudication (11).

ASI Small Works Contract (2000)

- A headnote indicates that the form is intended for smaller works which do not warrant the use of a bill of quantities. It should not be used for complex operations where the ASI Building Contract would be more appropriate. However, Supplementary Conditions at the back of the Conditions provide for bills of quantities, and for the appointment of a quantity surveyor with a role.
- Contract documents are drawings, and a Specification or Schedule of Works. The Contractor is to supply a full Schedule of Rates (1·4).
- Details relating to the particular contract are to be entered at relevant places in the text of the Conditions.
- The person appointed by the Employer as Contract Administrator is referred to as 'Architect'. If the person is 'Surveyor/Engineer' then this will be a Recital entry, and the wording of the Conditions is deemed to be changed.
- The obligations of the Contractor and the role of the Architect are listed (1 and 2).
- Key contract dates are entered in the Conditions (4·1), and the Architect may award at his or her absolute discretion 'an appropriate extension of time if warranted by the circumstances' (4·4).
- The contract requires Practical Completion, and this is defined (4·2).
- Failure to meet the completion date will result in payment of liquidated damages, as certified by the Architect (4·5).
- Domestic sub-contracting is permitted, but only to a limited extent, and subject to approval by the Architect (5·1).
- There is provision for nominated sub-contractors and nominated suppliers, subject to stated conditions (5·3 and 5·4).

ASI Forms

- The Contract Sum is VAT-exclusive. Fluctuations in scheduled rates and prices for labour and materials may be included (7·4); otherwise only changes in statutory charges will be allowable (7·3).
- The Architect may order variations, and the order will also stipulate the basis for valuation (i.e. according to the priced document; or rates approved by the Architect; or quotation approved by the Architect; or as daywork) (7·2).
- Architect's Interim Certificates are normally issued at monthly intervals (6·2) with payment by the Employer due within 14 days (6·3). Amounts for nominated sub-contractors must be identified in the Certificates and the firms advised (6·8).
- Architect's Final Certificate is to be issued after any defects have been made good, following the end of the Defects Liability Period – normally six months (6·9).
- The Contractor must comply with all legal and statutory requirements, including conforming with health and safety regulations (1·3 and 1·6).
- Insurance of new work in joint names against risk of damage by listed perils is by the Contractor (8·1), although the Employer may elect to carry this insurance. Insurance of existing buildings against risk of damage by listed perils is at the 'sole risk' of the Employer (8·2).
- The contract provides for determination of the employment of the Contractor for listed reasons of default on the part of either the Employer or the Contractor (9·1 and 9·2). There is a very limited reference to action and claims subsequent to determination.
- Disputes are to be submitted to the Architect initially for his or her decision (10·1). If not satisfactorily resolved they may proceed to arbitration (10·2). There is also a section on the right of parties to refer a dispute to adjudication, and the procedures relating to the appointment of an adjudicator and conduct of the adjudication (11).

ASI Minor Works Contract (2000)

- A headnote indicates that this form is intended for 'minor' works on a lump sum basis. No bills of quantities and no nominated sub-contractors.
- Contract documents are drawings and a Specification or Schedule of Work, and Health and Safety Plan.
- Details relating to the particular contract are to be entered at relevant places in the text of the Conditions.
- The person appointed by the Employer as Contract Administrator is referred to as the 'Adviser'.
- The obligations of the Contractor (1) and the role of the Adviser (2) are briefly described.

ASI Forms

- Key contract dates are to be entered in the Conditions (4·1). The Adviser may award, in his absolute discretion, an extension of time if the circumstances warrant it.
- The contract requires Practical Completion to be approved by the Adviser, and followed by a defects period – normally of six months (4·2).
- Failure to meet the completion date will result in payment of liquidated damages (4·1).
- Sub-letting is not permitted without the written consent of the Adviser (5·1).
- The Contract Sum is VAT-exclusive (A·3). The contract is fixed price except for changes in statutory payments (7·3).
- The Adviser may order variations, and the order will also stipulate the basis for valuation (7·1).
- Adviser's Interim Certificates will be issued from time to time at the discretion of the Adviser (6·2), with payment by the Employer due within 14 days (6·3).
- Adviser's Final Certificate will be issued after proper completion of the work and making good the defects after the end of the defects period (6·5).
- The Contractor must comply with all statutory requirements and pay all fees and charges legally demandable (1·3).
- Insurance of new work in joint names against specified perils is by the Contractor (8·1). Insurance of existing buildings against specified perils is at the 'sole risk' of the Employer (8·2).
- The contract provides for determination of the employment of the Contractor for listed reasons on the part of the Employer or the Contractor. There is very terse reference to actions which might follow (9·1 and 9·2).
- Claims or disputes are to be submitted to the Adviser initially for his decision (10·1). If not resolved, they may proceed to arbitration (10·2). There is also a section on the right of parties to refer a dispute to adjudication and the procedures relating to the appointment of an adjudicator and conduct of the adjudication (11).

7 Traditional procurement: shorter lump sum forms

ASI Forms

This contract?

If considering using ASI forms remember that:

The intended scope of the contract is carried in a headnote on each of the forms. They are all for lump sum contracts, and for private or local authority use. They are obviously for use by ASI members, but non-members might also find them of considerable interest and may use them.

The Building Contract and Small Works Contract both state that the law of the contract is English law, but the Minor Works Contract is silent on this point. All three forms refer to adjudication procedures being subject to the law of England and Wales. The forms would therefore not appear to be suitable for use under Scots law.

Architects should first be certain that the provisions adequately cover the nature of the intended work, and should check in particular that the provisions for certificates, insurance, determination and adjudication satisfy their requirements. They are somewhat limited, and probably best used for traditional work. Certainly as far as payment is concerned the forms seem to be light on detail required by the Housing Grants, Construction and Regeneration Act 1996 and might be subject to the Scheme for Construction Contracts. There is no provision for design by the Contractor, but nomination of sub-contractors and suppliers is provided for except in the Minor Works Contract. ASI sub-contracts are available for nominated and domestic sub-contracts.

Completing the forms is straightforward, with details either entered in the text of the clauses, or in the case of the Building Contract, in A5 of the Agreement.

The Contract Administrator is given considerable authority, and contract administration should prove a straightforward and conventional operation. There is a full range of ASI standard contract administration forms available.

Three compact and interesting forms, traditional in scope, clearly set out and structured, and plainly worded. Likely to appeal to clients, and certainly to smaller building firms.

Related matters

Documents

ASI Building Contract 2000 Edition
ASI Small Works Contract 2000 Edition
ASI Minor Works Contract 2000 Edition
ASI Domestic Sub-Contract 1996 Edition
ASI Nominated Sub-Contract 1981 Edition
ASI Contract Administration Forms
(eight in number, available in pad form)
ASI Agreement for Appointment of Professional Adviser
ASI Agreement for Appointment of Planning Supervisor

Notes

A Guide to ASI Contract Documents
(ASI publication)

SBCC Documents

Scottish Building Contract Committee (SBCC)

SBCC Forms of Contract

Background

Scotland might form part of the United Kingdom but Scots law, for historical reasons, differs markedly from the law of England and Wales. Not all statutes from Westminster apply to Scotland, and some legislation applies only to Scotland. Sources of law in Scotland are not the same as those of England, and the influence of Roman law and Canon law over time has resulted in a unique legal system and courts structure, not to mention a distinctive terminology. Close continental links in the past have also helped to produce what one commentator has described as an 'intermediate between Civil law and Common law'.

Despite pressure from England to conform to English law, the Scots have in large measure maintained their noble and independent ways. Therefore architects carrying out work in Scotland need to take account of the significant differences in legislation, and in the law relating to property, delict (broadly speaking the equivalent of the English law of tort) and of course contract.

Until fairly recently, building contracts in Scotland were often based on an exchange of letters rather than formal documents, and were direct trades contracts. The Emmerson Report of 1962 highlighted the different circumstances in Scotland and concluded that the existing trade by trade basis was unsuitable for main contracting in the building industry. It recommended that closer links should be established with London, and that a working party be appointed in Scotland, which duly resulted in the McEwan Younger Committee. A report from that committee, with the initiative of the Royal Incorporation of Architects in Scotland, resulted in the formation of the Scottish Building Contract Committee (SBCC) in 1964.

The SBCC comprises representatives from:

- Royal Incorporation of Architects in Scotland;
- Scottish Building Employers' Federation;
- The Royal Institution of Chartered Surveyors in Scotland;
- Scottish CASEC;
- Convention of Scottish Local Authorities;
- National Specialists Contractor's Council;
- Association of Consulting Engineers (Scottish Group);
- Scottish Executive, Building Division;
- Confederation of British Industry;
- Association of Scottish Chambers of Commerce.

SBCC Documents

This is obviously a wide representation compared, for example, with the JCT Ltd.

The Banwell Report, issued in March 1964, included a recommendation that 'a common form of contract for all construction work covering England, Scotland and Wales was both desirable and practicable'. Implementation of this report resulted in the Joint Contracts Tribunal inviting the Scottish Building Contracts Committee to join the Tribunal, and this invitation was accepted in 1966 'subject to the right of the SBCC to issue Tribunal forms in a manner which conformed with Scots law and practice'.

The SBCC is therefore primarily responsible for the adaptation of JCT documents for use in Scotland. It is also responsible for amending and publishing such forms, advising on the interpretation of these contracts, promoting best practice, drafting and publishing guidance and practice notes, nominating arbiters, mediators or third party tribunals, and attending national and other committees including those at the JCT. The SBCC is now the 'Relevant College' for the Scottish Construction Industry within JCT Ltd, and takes an active part in providing input to drafting where Scots law and practice dictate.

Nature

The SBCC Building Contracts are referred to as Scottish Supplements, but they are quite distinct and incorporate only those Conditions from the JCT forms which are relevant. For example:

Scottish Building Contract With Quantities (2002 Revision) with Scottish Supplement *(Private, Local Authorities, Fluctuations, Formula Rules, Notes to Users)*;

- Scottish Building Contract Without Quantities (2002 Revision) with Scottish Supplement (Private, Local Authorities, Fluctuations, Formula Rules, Notes to Users);
- Scottish Building Contract With Approximate Quantities (2002 Revision) with Scottish Supplement (Private, Local Authorities, Fluctuations, Formula Rules, Notes to Users);
- Scottish Building Contract Sectional Completion Editions (2002 Revision) with Scottish Supplement (With Quantities, Without Quantities, With Approximate Quantities, Fluctuations, Formula Rules, Notes to Users);
- Scottish Building Contract Contractor's Designed Portion Editions (2002 Revision) with Scottish Supplement (With Quantities, Without Quantities, Fluctuations, Formula Rules, Notes to Users);
- Scottish Building Contract Contractor's Designed Portion and Sectional Completion Editions (2002 Revision) with Scottish Supplement (With Quantities, Without Quantities, Fluctuations, Formula Rules, Notes to Users);
- Scottish Measured Term Contract for Maintenance and Minor Works (2002 Revision) with Scottish Supplement (Notes to Users);

SBCC Documents

- Scottish Management Contract (2002 Revision) with Scottish Supplement;
- Scottish Management Contract Phased Completion Edition (2002 Revision) with Scottish Supplement.

(Both these Management Contracts need the Works Contract/1/Scot (2002 Revision) Invitation; Tender and Works Contract; Scottish Employer/Works Contractor Agreement; Works Contract/3/Scot (2002 Revision); Contract of Purchase from a Works Contractor (2002 Revision) and SBCC Notes to Users.)

In addition the SBCC also publishes its own versions of nominated and domestic sub-contract forms, the SBCC Contracts of Purchase, SBCC versions of consumer contracts, and the Scottish Minor Works Contract (2002 Edition).

Use

The Conditions of the adapted forms are largely those of the JCT documents. Modifications relate to definitions, nominated sub-contractors and suppliers, determination, payment for off-site goods and materials, deducting liquidate damages etc.

The most significant differences lie in the fact that the modified or adapted JCT Conditions are incorporated by the Scottish Supplement, and that attestation is to be made in accordance with Scots law. The SBCC issues a note on attestation procedures. Another difference concerns dispute resolution in Scotland. The Arbitration Act 1966 does not apply to Scotland, and arbitration procedures and conduct are quite distinct. Adjudication is a statutory right and the SBCC has produced its own Adjudication Agreement.

Architects undertaking work in Scotland who are not familiar with Scots law and the SBCC Contracts would do well to seek advice from the SBCC.

Synopsis of SBC MW and comparisons with JCT MW98

The SBCC Scottish Minor Works Contract (January 2002 Edition), although copyright of SBCC, is clearly closely modelled on the JCT MW98. The differences in the Conditions between the two forms (described as an Agreement in the case of MW98, and a Contract in the case of the SBC MW) may be summarised as follows.

1 Intentions

- The wording of the clauses in this section is identical, though not the sequence.
- Applicable law (clause 1·7 in MW98) is stated to be the law of Scotland (11).
- Third party rights (clause 1·8 in MW98) are not referred to as the Contracts (Rights of Third Parties) Act 1999 does not apply to Scotland.

SBCC Documents

2 Time

- Covered under section heading 2 in MW98, but under section heading 3 in SBC MW due to a provision for bills of quantities in the latter under section heading 2.
- There is provision of damages for 'late completion' (3·3), arguably a term more accurate than the 'non-completion' used for MW98 clause 2·3.

3 Control

- Covered under section heading 3 in MW98, but SBC MW separates into section heading 4 Assignation and Sub-Contracting, and section 5 Control of the Works.
- Wording of the provisions is mainly the same in both forms, but SBC MW provides for listed sub-contractors.
- The main contractor may chose a domestic sub-contractor from a list of not less than three names as set out in the Contract Documents. There is a mechanism for maintaining the list at three, and a fall-back provision where this proves impossible (4·3 and 4·4). This is similar to the JCT98 clause 19·3.

4 Money

- Covered under section heading 4 in MW98, but section heading 6 in SBC MW.
- The wording of the clauses is identical in both forms with the exception of the fixed price provision (4·7 in MW98) (6·7 in SBC MW) where although the wording differs, the content appears to be the same.

5 Statutory obligations

- Covered under section heading 5 in MW98, but section heading 7 in SBC MW. The wording of the clauses is identical in both forms, with the exception of that concerning prevention of corruption. The SBC form takes account of Scottish legislation (7·5).

6 Insurance

- Covered under section heading 6 in MW98 but section heading 8 in SBC MW. The wording of the clauses is identical in both forms, except for a minor difference in the last lines of SBC MW clause 8·3A.

7 Termination

- Covered under section heading 7 in MW98 but section heading 9 in SBC MW. The wording of the provisions is identical in both forms, except for adjustments to take account of Scottish legislation (9·1·2 and 9·2·2).

8 Miscellaneous

- Supplemental Conditions are the same for both forms in respect of A: Contribution Levy and Tax Changes; B: Value Added Tax; and C: Construction Industry Scheme.

SBCC Documents

- Unlike MW98, the SBC MW brings the provisions for adjudication, arbitration and court proceedings into the main body of the Conditions rather than treat them as Supplemental.
- SBC MW, unlike MW98, does not introduce attestation at the front of the form, but has a section heading 13: Registration. In a separate Note to Users: Attestation, the SBCC offers clear advice on attestation procedures under Scots law, including examples of appropriate wording for testing clauses.

9 Disputes

- Part II of the Housing Grants, Construction and Regeneration Act 1996 applies to Scotland, and either party has a statutory right to refer any difference or dispute arising under the contract for adjudication.
- Unlike MW98, SBC MW has no article referring to dispute resolution, but the 'Considering' preambles refer to clauses 10A: Adjudication, 10B: Arbitration and 10C: Court proceedings.
- The procedures and rules for adjudication in clause 10A differ in wording from those in MW98, and SBCC has its own Adjudication Agreement.
- Arbitration in Scotland is not subject to the Arbitration Act 1996, and clause 10B sets out the relevant procedures and Codes which will apply. This will apply unless it has been deleted.
- The alternative of court proceedings is referred to in clause 10C.

SBCC Documents

This contract?

If considering using SBC MW remember that:

It is intended for building works of a simple straightforward nature, and is not suitable for work of such duration that full fluctuations provisions are required. It can only be used where the Employer has engaged a professional consultant to act as Contract Administrator. There is provision for bills of quantities, and a quantity surveyor may be appointed, although he or she appears to have no specified role under the contract for such matters as valuations.

This contract is the only standard form for minor building works in Scotland. The form is drafted to ensure compliance with the CDM Regulations, and with Part II of the Housing Grants, Construction and Regeneration Act 1996.

As with JCT MW98, there is no provision for phased completion, fluctuations, or design by the Contractor. Unlike the JCT form, this SBC contract provides for the naming of sub-contractors by way of a list of three names and there is also a dedicated SBC sub-contract available.

When completing the form, decisions are required relating to insurance of the Works and damages for late completion. Appropriate deletions will indicate whether the final resolution of disputes is by arbitration or court proceedings.

If acting as Contract Administrator, the Conditions are likely to prove adequate for most situations. The procedural rules are simple.

Unlike the JCT MW98, this SBC form can be seen as more of a middle range contract for all work where use of the full SBC version of JCT98 cannot be justified. There is no SBC equivalent of the JCT IFC98 form.

Related matters

Documents

SBCC Scottish Minor Works Contract January 2002 Edition
SBCC Scottish Minor Works Sub-Contract January 2002 Edition

Notes

SBCC Notes to Users: Attestation
SBCC Guidance Notes upon Dispute Resolution in Scotland

Traditional procurement 8

Consumer contracts

The Joint Contracts Tribunal Ltd
Agreement for Housing Grant Works HG(A) 2002 Edition

The Joint Contracts Tribunal Ltd
Building Contract for a Home Owner/Occupier

The Joint Contracts Tribunal Ltd
Contract for Home Repairs and Maintenance

Architecture and Surveying Institute
Mini Form of Contract 1998 Edition

These are contracts between 'consumers' (i.e. a natural person entering into a contract for purposes outside business, for example a person wishing to have work carried out on his or her own house) and a 'contractor' (i.e. a supplier of goods or services in the course of their business).

The law seeks to protect consumers so that they are not put at a disadvantage. The Unfair Terms in Consumer Contracts Regulations 1994 (SI 1994/3159) apply to contracts which have not been individually negotiated (e.g. this would generally include all standard forms of building contract and those for professional services), and require that there must be no unfair terms. Terms which are deemed to be unfair are those which might cause a significant imbalance in the parties' rights to the detriment of the consumer and such terms will not be binding on the consumer. The Regulations require contracts to be expressed in plain, intelligible language.

The forms in this chapter will normally be classed as consumer contracts, although some might also have a wider use.

JCT HG(A)

The Joint Contracts Tribunal Ltd

Agreement for Housing Grant Works HG(A) 2002 Edition

Background

In 1970 the Joint Contracts Tribunal first produced a variant of the Agreement for Minor Works 1963 Edition, for use with housing contracts which were grant-aided by the local authority. Renovation grants made under the Local Government and Housing Act 1989 saw the introduction by the JCT of the 1994 Agreement for Renovation Grant Works, which was produced in two versions. The white-covered RG(A) closely followed the wording of the 1980 Minor Works form and was for use where an architect was appointed. The pink-covered RG(C) was a more maverick form and was for use where the client dealt directly with the Contractor and no architect was appointed.

With the introduction of the Housing Grants, Construction and Regeneration Act 1996, a revised form of contract became necessary and the Agreement for Housing Grant Works HG(A) was published in 2002. It supersedes the RG(A) 1994 Edition. The contract is for use where an Architect or Contract Administrator is appointed, and a grant is to be made under Part I of the Housing Grants, Construction and Regeneration Act 1996. There are apparently no plans to revise the version for use where the client deals directly with the Contractor without appointing an Architect or Contract Administrator, and the former RG(C) is now so outdated that it must be regarded as dead. Also, of course, newer consumer contracts specially drafted for use in such situations have become available.

Nature

The 2002 Edition of Agreement for Housing Grant Works is closely modelled on the Agreement for Minor Building Works MW98. The total document runs to 31 pages but the Conditions take only 10 of these. The Agreement includes six Recitals and seven Articles, and the Conditions are section headed. An Appendix comprises copies of a form of acknowledgement and authority which must be completed by or on behalf of the payee of the grant, and sent to the local authority. This sanctions payments by the local authority direct to the Contractor.

A very helpful Guidance Note is included with the form, and this precisely summarises the types of grant currently available, as well as giving some worthwhile advice on contract administration relating to grant aid.

The Conditions reflect the nature of the work, and the wording is appropriate for

JCT HG(A)

'construction contracts' to which Part II of the Housing Grants, Construction and Regeneration Act 1996 do not apply. Where Part II does apply, then particularly concerning notices over payment procedures, the statutory obligations must obviously be complied with in addition to the express requirements in the contract wording.

Adjudication is included as probably the most suitable way of resolving disputes, with the parties entitled to refer final resolution to litigation. If the parties, for some good reason, still insist on going to arbitration, then suitable arbitration provisions would need to be incorporated at the outset, and those to be found in MW98 would probably be suitable.

Use

The form is intended for use where an Architect or Contract Administrator has been appointed, and where a grant is to be made under Part I of the Housing Grants, Construction and Regeneration Act 1996.

The Works are to be carried out for an agreed lump sum, with no provision for fluctuations – not even tax etc. changes. The contract documents may comprise contract drawings, contract Specification, and or Schedules. The Contractor's price will be lump sum, based upon pricing of the relevant contract documents.

The Fourth Recital states that the client Employer has applied for a grant, and the Fifth Recital states the sum for which a grant has been approved and what arrangements are made for payments of the grant.

The form provides for executing under hand and not as a deed, or alternatively as a deed.

Synopsis

1 Intentions

- The Contractor is obliged to carry out and complete the Works in accordance with the contract documents and the Health and Safety Plan where applicable (1·1).
- The Contractor is to perform with due diligence and in a good and workmanlike manner (1·1).
- The quality and standards are those described in the contract documents (1·1).
- The Architect is obliged to issue any further information necessary to enable the Contractor to carry out the Works (1·2).
- Inconsistencies in or between contract documents shall be corrected and this may be treated as a variation instruction (3·6).
- The printed Conditions prevail (3·6).

JCT HG(A)

- Where all the CDM Regulations apply, a Planning Supervisor and Principal Contractor must be identified, and a reappointment made if necessary (Sixth Recital, Articles 4 and 5 and clauses 1·3 and 1·4). Where CDM Regulations 7 and 13 only apply, the Contractor is to notify the Health and Safety Executive before carrying out work.

2 Time

- The Works may be commenced on a date to be inserted and shall be completed by a date to be inserted (2·1).
- The Contractor is to notify the Architect if completion by the stated date is unlikely, for reasons which are not within the Contractor's control (2·2). Difficulties with sub-contractors or suppliers are within the Contractor's control.
- The Architect is empowered to make such extension of time as may be reasonable. (Note: no specific grounds are listed.)
- If the Contractor fails to complete by the date, the Employer is entitled to liquidated damages (2·3) and the rate is to be inserted. (Note: there is no reference to a certificate of non-completion.)
- Practical Completion is certified by the Architect (2·4).
- Following this the Contractor is obliged to rectify defects. (The Defects Liability Period is three months, although a longer period may be inserted, 2·5). A certificate is to be issued by the Architect when this obligation has been discharged.
- There is no provision for deferring possession, nor for partial possession, nor for sectional completion.

3 Control

- The bar to assignment without written consent refers to 'this Agreement' (3·1). Written consent to sub-contract is required (3·2).
- There is no provision for naming a sub-contractor, and no provision for including a list of approved firms.
- Architect's instructions are to be issued in writing (3·5) but may be oral and confirmed in writing. Instructions may include variations (3·7) and this would presumably allow for postponement, as no specific reference is made elsewhere.
- The Contractor's 'competent person' is to be on site at all reasonable times (not constantly) (3·3). There is no provision for a Clerk of Works.
- There is no provision for testing, opening up or action in the event of non-compliance. Non-payment would seem to be the ultimate sanction.
- There is no provision for work to be carried out by persons engaged directly by the Employer.

JCT HG(A)

4 Money

- The Contract Sum is VAT-exclusive (Article 2).
- The contract is fixed price (4·5) unless stated otherwise in the contract documents.
- Where a variation incurs additional work, the Architect must advise the contractor on whether this will be eligible for grant aid (3·7).
- Instructions must be given where provisional sums are included (3·8). The cost of such work is either to be agreed in advance or valued as for variations.
- Direct loss and/or expense is limited to variations or provisional sum work, and included in the valuation (3·7).
- Provided that the contract period is longer than 44 days, the contractor can request that progress payments are made at four-weekly intervals (4·1·1). The Architect is to issue certificates to the Employer. Retention will be at 5 per cent (unless a different figure is entered).
- Certificates will show the amount due and the basis of calculations. The final date for payment by the Employer is 14 days from issue of the certificate. Failure to pay by the final date for payment will attract interest of 5 per cent over the current base rate, and can also give the Contractor a valid right to suspend work (4·6).
- A penultimate certificate is to be issued within 14 days after Practical Completion releasing all outstanding money except half the retention (4·2).
- The Final Certificate is to be issued within 28 days of either issue of the certificate signifying completion of making good the defects, or the Contractor submitting all necessary information for the final account (4·3). He is given three months (unless a different figure is inserted) from the date of Practical Completion to produce this.
- For the Final Certificate, the final date for payment is 14 days from the date of issue of the Certificate.
- Where the Employer is not a residential occupier (for example in the case of a landlord with a houses in multiple occupation grant) then JCT HG(A) might be a 'construction contract' to which Part II of the Housing Grants, Construction and Regeneration Act 1996 applies. In this case the Employer would be obliged to give notices to the Contractor as required under sections 110 and 11 of the Act. Notices would have to be given within the applicable time limits, stating the amount proposed to be paid, and any intention to withhold or deduct money. This would apply to Interim, penultimate and Final Certificates, despite the absence of any such express provision in the contract Conditions.
- Clause 4.4 covers the payment of the grant or any part of it directly to the Contractor by the local authority. Signed copies of the form of acknowledgement and authority

(in Appendix to JCT HG(A)) must have been sent to the local authority and Contractor by the Employer. Certificates issued by the Architect will then indicate the amount of the grant to be paid, and this amount should be deducted from the amount otherwise due to be paid by the Employer. This balance shall be paid by the Employer within 14 days of the date of issue of the Certificate.

5 Statutory obligations

- It is the Contractor's duty to comply with all statutory obligations, including giving all required notices (5·1).
- The Contractor is to notify the Architect in writing if he finds any conflict between statutory requirements and his other contractual obligations (5·1). Having done that, he will not be liable to the Employer under this contract.
- Where the CDM Regulations apply in full and Alternative B in the Sixth Recital applies, the Employer is to ensure that the Planning Supervisor and Principal Contractor carry out the duties required of them under the CDM Regulations.

6 Insurance

- The Contractor indemnifies the Employer in respect of personal injury or damage to property. He is required to arrange insurance to back this (6·1 and 6·2). The minimum cover as a contractual obligation requires an entry (6·2).
- The Contractor is obliged to arrange insurance of new works in joint names against damage by perils specified in clause 6·3A. This is for full reinstatement value, but note that these perils do not specifically include theft, vandalism or impact, and advice should be sought from insurers. A percentage to cover professional fees should be inserted.
- Joint names insurance of existing structures, together with any new work which forms part of an alteration or conversion, is to be arranged by the Employer. As with clause 6A the contractual requirement is for cover only against specified perils (6·3B). In the event of loss or damage the Architect must issue instructions, and the cost of making good is to be valued under clause 3·7 as for a variation.

7 Termination

- The Employer may determine the employment of the Contractor if he fails to proceed diligently with the Works, or wholly or substantially suspends carrying out the Works, or if he becomes insolvent (7·2).The latter does not mean automatic determination.
- In such circumstances the Architect may issue a warning notice, but notice of actual determination is a matter for the Employer.
- The Contractor may determine his own employment for stated reasons (7·3).
- There is no provision for determination by either party for neutral causes.

JCT HG(A)

- The respective rights and duties of the parties following determination are set out in the Conditions (7·2 and 7·3).

8 Miscellaneous

- There is no definitions clause.
- There is no reference to access for the Architect – presumably this is implied.
- The Architect may order the exclusion of any person from the Works (3·4).
- There is no reference to antiquities, terrorism cover, etc.
- There is a contracting out of third party rights under the Contracts (Rights of Third Parties) Act 1999 (1·8).

9 Disputes

- Article 6 of JCT HG(A)98 provides for adjudication.
- Procedures for referral to adjudication, and the appointing of an adjudicator, the powers and conduct of an adjudicator are set out in clause 8.
- The adjudicator's decision is binding on the parties, at least until the dispute is finally determined by legal proceedings.
- Article 7 determines that subject to adjudication, disputes or differences shall be finally decided by legal proceedings. There is no provision for arbitration.

JCT HG(A)

This contract?

If considering using JCT HG(A) remember that:

It is intended for use in housing improvement work, where a Contract Administrator has been appointed and a grant is to be made under the Housing Grants, Construction and Regeneration Act 1996. It is for lump sum contracts of relatively short duration, with a fixed price. There is no provision for the appointment of a quantity surveyor.

The form is for use only in England and Wales. Part I of the Housing Grants, Construction and Regeneration Act 1996 does not extend to Northern Ireland or Scotland.

There is reference to grant payment procedures in Fourth and Fifth Recitals, and clause 4·4. The Appendix is a form of acknowledgment and authority for grant payment. There is provision for adjudication, but not for arbitration. Litigation is the final means of dispute resolution.

When completing the form, decisions are required relating to damages for non-completion and insurance of the Works. Where the contract is one to which Part II of the Housing Grants, Construction and Regeneration Act 1996 applies, some modification of the payment clauses will be required (as those for MW98).

If acting as Contract Administrator the procedural rules are brief and relatively easy to administer. Timescales are relaxed, but it is important to watch out for any grant stipulation that work must be completed within a specified time. The RIBA does not publish contract administration forms for HG(A), but some of those for MW98 could be applicable.

The form is a close relative of JCT MW98. A helpful guidance note is included, which summarises the types of grant currently available and provides commentary on the Conditions.

Related matters

Documents

Agreement for Housing Grant Works HG(A) 2002 Edition

References

JCT Guidance Note Housing Grant Works (included with form of contract)

JCT Guidance Note for MW98 (included with form of contract)

Commentaries (relating to MW98 but relevant)

David Chappell and Vincent Powell-Smith
The JCT Minor Works Form of Contract
2nd edn, Blackwell Science (1998)

Sarah Lupton
Guide to MW98
RIBA Publications (1999)

JCT Home Owner/Occupier Contracts

The Joint Contracts Tribunal Ltd

Building Contract for a Home Owner /Occupier

Building Contract and Consultancy Agreement for a Home Owner /Occupier

Background

It used to be thought that whilst home improvements might prove lucrative for smaller builders, such work for the most part was not of interest to professional consultants. However, the recent boom in housing improvements and the potential for engagement in this market have shown the need for forms of contract appropriate to this kind of work.

The JCT has produced two forms of building contract for home owners or occupiers intending to carry out domestic building work. They both follow a relatively innovative pattern in that they are packaged in a folder, there are counterpart copies of the forms for customer and the builder, and guidance notes. They are written in a refreshing style commendably free from jargon and legalist language which has gained for them the Crystal Mark for Clarity approved by the Plain English Campaign. An attempt has been made to market them widely through retail outlets such as book shops directly to customers.

The first form published in 1999 was the Building Contract for a Home Owner/ Occupier, suitable for use where the customer chose to deal directly with a builder for home improvements, small extensions or repairs. Obviously this promised a significant improvement over oral arrangements or poorly worded letters, and encouraged the parties to consider methodically and agree the important points before concluding the deal.

The second form published in 2001 was the Building Contract and Consultancy Agreement for a Home Owner/Occupier who has appointed a consultant, and it followed the same attractive format for packaging and presentation. This could prove to be a more appropriate form for much of the larger-scale domestic work now attracting the involvement of architects and building surveyors, than the JCT Minor Works Contract which is perhaps more suited to smaller work of a commercial nature. It can be used only where a consultant is appointed to act on the customer's behalf,

JCT Home Owner/Occupier Contracts

and the document folder includes a JCT Consultancy Agreement for professional services which is dedicated for use with this building contract. It is likely to prove more appropriate in these circumstances than agreements published by the professional bodies.

Nature

The Building Contract for a Home Owner/Occupier is just seven pages long, and is in two parts. First there is a questionnaire in duplicate which when completed sets out the arrangements for the work. Secondly there are the contract Conditions clearly and simply worded under 11 headings (also two copies). The customer is provided with an enquiry letter to send to a potential builder, and a sheet of helpful guidance notes.

The Building Contract and Consultancy Agreement for a Home Owner/Occupier is rather more formal in appearance. There is a Building Contract which has two parts; Part 1 is the arrangements for the work, and Part 2 is the Conditions set out under 13 headings. The Consultancy Agreement is also a two part document; Part 1 is the consultant's services and Part 2 is the Conditions set out under nine headings, a total of just 14 pages of text. There are copies of each for the customer and the Contractor, and Guidance Notes are printed on the inside cover of the package.

Use

Both forms are simple and easy to understand. They are only suitable for construction contracts to which Part II of the Housing Grants, Construction and Regeneration Act 1996 Act 1996 does not apply. The CDM Regulations will not apply.

The earlier Building Contract for a Home Owner/Occupier must not be used where a consultant is employed. There is no stated cost limit, but as will be obvious from the excellent tick box approach used for the arrangements document, and the basic contract Conditions, it is for very small domestic work only. Incidentally, for architects who are involved in a partial service for domestic work but whose clients nevertheless ask for advice about appointing a builder, this form would be a sound recommendation.

The Building Contract and Consultancy Agreement is for a home owner/occupier who intends to appoint a consultant to deal directly with the builder. Note that it is suitable only for a home owner and if the work is being carried on as a business venture, then one of the commercial contracts should be used.

Brief synopsis

1 Building Contract for a Home Owner/Occupier 1999

- Lump sum contract. Note that Contractor's quotation is VAT-inclusive.
- Drawings and/or Specification or other documents. No bills of quantities.
- Either customer or Contractor may deal with applications for planning permission,

JCT Home Owner/Occupier Contracts

building regulations and party wall consents. (No mention of fees.)

- Facilities for Contractor, as ticked, to be provided free of charge by the customer.
- Contractor's responsibilities include storing equipment at end of each day, regularly disposing of rubbish, and leaving all clean and tidy after finishing the work.
- Payment may be single payment or by agreed instalments. Customer to pay on invoices within 14 days, 95 per cent of amounts due.
- Working period may be by agreed duration, or start and completion date entered. Working hours may be specified.
- Only customer can change work details. Contractor to submit a price before work goes ahead.
- Customer can extend work period if Contractor is delayed for limited stated reasons.
- Insurance provisions limited to customer's household insurance, Contractor's all risks cover for damage to work and unfixed materials, and public liability cover.
- Provision for occupation and security of premises, very limited.
- Either customer or Contractor can bring contract to an end for stated reasons.
- Defects period is three months, and customer will pay remaining 5 per cent of money due within 14 days of Contractor putting right any faults promptly reported by the customer.
- If disputes arise, either party can start court proceedings, or may opt for adjudication.

2 **Building Contract and Consultancy Agreement for a Home Owner/Occupier 2001**

- Description of the work is detailed in Contractor's quotation/consultant's drawings /consultant's specification. No bills of quantities, but 'other documents' may be included.
- Lump sum contract. Contractor's price is VAT-inclusive. Contractor to itemise this and show details of VAT chargeable.
- Price includes for 'unexpected problems' which should have been foreseen by the Contractor from the documents or a site visit. Contract figure can only be changed up or down if changes to the work are instructed.
- A start and completion date should be entered, or a contract period stated. Working hours may be specified.
- The customer can expect to receive the benefit of any product guarantees.
- The Contractor's responsibilities include carrying out work 'as the consultant instructs' and to be 'at the premises regularly to carry out the work' and to 'keep to all his legal duties and responsibilities'.

JCT Home Owner/Occupier Contracts

- The customer's responsibilities include providing access to the premises throughout the working period, keeping working areas clear of obstructions, and allowing the Contractor to carry out work in an order which he considers necessary to complete on time.
- The consultant will act for the customer in giving instructions, extensions of time, and in issuing certificates. There are only two certificates, first on completion of the work, and second after the Contractor has rectified faults at the end of the defects period of three months. Although there are no monetary certificates, the consultant must be satisfied that the Contractor's invoices are correct. The contract states that in the event of any disagreement with the consultant's decisions, the Contractor must take this up directly with the customer.
- Changing the work details can be ordered only by the consultant. Where an increase in work is likely, the Contractor must quote the extra cost and time involved before work can be authorised. Where change amounts to a reduction of work, the Contractor will make an appropriate deduction.
- The consultant can make a fair and reasonable extension if the Contractor is unable to complete to time for reasons beyond his control. If delay is caused by the consultant or the customer the Contractor may also be entitled to costs.
- Consultant is responsible for planning permission, building regulations and party wall consents. (Consultants acting for the customer in serving party wall notices should make sure that they have written authorisation, otherwise their capacity to act might be challenged.)
- Welfare facilities to be provided for Contractor free of charge by customer are as ticked.
- The Contractor is to take all practical steps to prevent or minimise health and safety risks to the customer and other occupants, to minimise environmental disturbance, nuisance or pollution. In return the customer undertakes to take notice of warnings by the Contractor and not knowingly allow occupants or visitors to be exposed to dangers.
- Payment may be a single payment or by instalments as agreed. Customer to pay 95 per cent of total amount due. If stage payments are selected, they should be clearly defined and amounts stated.
- Customer is expected to pay 95 per cent within 14 days after completion has been certified by the consultant. There is a three months Defects Liability Period, at the end of which the consultant will issue a list of any faults. When these have been rectified the remaining 5 per cent will be due within 14 days after completion has been certified.
- Insurance provisions are minimal, and limited to householder's insurance, contractor's all risks cover for full cost of damage to work and unfixed materials (no perils stipulated) and public liability cover. The contract liability cover figure is to be entered.

JCT Home Owner/Occupier Contracts

- Provisions for occupation and security of premises are limited. If remaining in residence the customer should be made aware of the inconvenience and disruption which might result, and the Contractor should allow for temporary protective measures as appropriate. If the customer is vacating the premises, house insurance policies should be checked, as most suspend cover where premises are left unoccupied beyond a stated period and special arrangements might become advisable. Also the Contractor's obligation to adopt 'practical and common-sense precautions' might require something more specific in the Specification. The CDM Regulations are unlikely to apply in full to work of this nature.
- If disputes arise, either party can start court proceedings or may opt for adjudication. In the latter event the Contractor may not apply to the National Specialist Contractors Council to be appointor.
- Other rights and remedies are not extinguished by the contract provisions.

JCT Home Owner/Occupier Contracts

This contract?

If considering using JCT Home Owners Contracts remember that:

Each form is intended for use only with work for home owners/occupiers. If the customer wishes to appoint a consultant, then the only one to be used is the version which gives a role for the consultant. In such cases the dedicated Consultancy Agreement must also be used. This document takes a very practical and clear approach to defining the services and the conditions which will apply. These differ considerably from the RIBA Small Works form (SW/99) which should not be used.

Neither of the forms is suitable for use in Scotland.

Completing the forms should be straightforward, but care is need to make sure that duplicate copies carry identical entries.

Problems could exist over possession and occupation, work sequence, design responsibility particularly relating to services installations or other specialist work, and insurance.

Helpful guidance notes are provided for each contract, and they should be read carefully, particularly in the case of the version where a consultant is appointed. As yet there are no dedicated contract administration forms available, but letters should be clearly identified as certificates where they are intended to serve as such.

Each of these forms adopts a completely fresh and friendly approach which will initially be unfamiliar to most contract administrators. The presentation and packaging is novel and customer orientated. The terminology is straightforward and the forms are commendably written in plain English. As Tony Bingham was moved to comment in *Building*, 5 November 1999, 'This little form is a beauty'.

Related matters

Documents

JCT Building Contract for a Home Owner/Occupier 1999
JCT Building Contract and Consultancy Agreement for a Home Owner/Occupier 2001

References

JCT Guidance Notes included for each of the forms

Home Repairs and Maintenance Contract

The Joint Contracts Tribunal Ltd

JCT Contract for Home Repairs and Maintenance

Background

The JCT is obviously trying to ensure that there is a form for all situations. This is a home care and repair simple contract published in 2002. It is unlikely to attract much attention from the construction professionals, who have no role under its terms.

Nevertheless it should provide a most useful service in bringing consumers and builders to think about, and commit to writing, those essentials of the contract which in the past have too often be left as vague assumptions.

Where the contractor is previously unknown to the customer, but quite ready to enter into an agreement such as this, it could also signify that this is probably a reputable firm.

Nature

This must be the shortest standard form for building work ever produced. The contract covers just three sides of an A5-sized triple fold document. There are counterpart copies for the customer and the Contractor, distinguished by the diagonal flash at the top corner of the cover. The Conditions are minimal but appropriate for this kind and size of operations, clearly worded and set out in 11 short sections. As with other JCT contracts for home owners, this one also comes with the Crystal Mark of the Plain English Campaign.

Use

The guidance notes suggest that repairs and maintenance might, for example, cover electrical rewiring, plumbing installations or even painting and decorating. The form is not intended for use with building work which might involve structural alterations, nor where more than one trade is likely to be involved.

An important point emphasied is that the customer should let the Contractor know at the time of requesting a quotation that this contract will be used. This will then reduce the likelihood of the Contractor submitting a quotation which is subject to the small print of his own terms.

Brief synopsis

- A two line description of the intended work is followed by a tick box indicating whether the premises will be in occupation at the time.

Home Repairs and Maintenance Contract

- Documents are a 'specification' prepared either by the customer or the Contractor, and the Contractor's quotation.
- The Contractor's price is inclusive of VAT, and may be a fixed lump sum or an hourly rate.
- Facilities to be provided by the customer free of charge are shown in tick boxes.
- Payment is not made until the work is completed, and within 14 days of receipt of the Contractor's itemised invoice.
- The Contractor must carry 'enough insurance' to cover full costs of damage and materials on site, in addition to public liability insurance.
- A start date and the contract period are entered, together with the agreed daily hours of working.
- The Contractor's obligation is to carry out the work 'competently and carefully', and to leave the work areas clean and tidy each day.
- Sub-contracting is permitted, but requires the customer's permission.
- The Contractor is responsible for health, safety and environmental matters.
- Third party rights are excluded, but the contract does not rule out other legal remedies open to either party.
- There is no provision for loss or expense, but the contract accepts that the parties may claim from each other in the event of failure to keep to the contract.
- Disputes can be referred to adjudication, and the right to litigation is preserved.

Home Repairs and Maintenance Contract

This contract?

If advising using JCT Home Repairs Contract remember that:

It is intended for repairs and maintenance work, to which Part II of the Housing Grants, Construction and Regeneration Act 1996 will not apply. The work will almost certainly not be subject to the full CDM Regulations.

It is not for use in Scotland.

It is a modest form, little more than a leaflet in appearance, but it must provide a more business-like way of dealing than just relying on a Contractor's estimate, whether written or oral. The intended repair works may be simple, but the cost could seem a relatively large expense to the customer and should be covered by a proper agreement.

There is no role under the contract for construction professionals, but they might be approached for advice on a suitable form and if so this should be a safe recommendation. It might also help to protect customers from the so-called 'cowboys' of the industry!

Related matters

Documents

JCT Contract for Home Repairs and Maintenance 2002

References

JCT Guidance Notes for customers included in the printed form

ASI Mini Forms

The Architecture and Surveying Institute

Mini Form of Contract (General Use)

Mini Form of Contract (Home Improvement Agencies)

Background

In 1998 the ASI took an innovative step and produced a document which they neatly styled a Mini form of contract. Before this it was impossible to find a standard form appropriate for very small jobs, which also provided for the client to appoint an Architect or other professional to act as a Contract Administrator. At the time the Mini had no real competitors.

The form is produced in two versions, one for general use (blue) and the other developed in collaboration with Care and Repair England for use with residential works (yellow).

Nature

In appearance and structure the forms are somewhat conventional, and compared with the latest JCT consumer contracts, not so friendly. Having said that, they are written in plain English and require the minimum of effort to complete.

Both versions of the form have four pages of Conditions, and come with a dedicated Letter of Invitation and Form of Tender. The Conditions are to a section headed format consistent with that adopted generally for ASI contracts.

Use

The forms are produced primarily for use by members of the ASI, but there is no bar to them being used by non-members of that Institute.

They are recommended for use in contracts up to the value of £6,000 at 1998 prices, and both require the Employer to appoint an 'Adviser' to act as Contract Administrator.

Whilst the form for general use might well constitute a construction contract to which Part II of the Housing Grants, Construction and Regeneration Act 1996 applies, the version for residential work would certainly not. Whether or not the CDM Regulations applied in full would depend on the nature, scale and duration of the intended works, but there is no express contractual obligation in the Conditions.

ASI Mini Forms

Brief synopsis

1 ASI Mini Form of Contract (General Use)

- Completed Form of Tender and Letter of Acceptance together with the 'contract documents' constitute the agreement between the parties.
- Contract Documents are referred to as Specification/Schedule of Works/drawings prepared by the Adviser. They are to be signed by both the parties.
- Contract Sum is VAT-exclusive. Any adjustments to the tender before acceptance will be entered in the Schedule of Adjustments.
- The Contractor's obligations include for executing the work 'to a standard of finish to the Adviser's satisfaction' (this might cause problems over conclusiveness of the Final Certificate) (1·1 and 6·4).
- The Adviser's role includes inspection of the Works, and the issue of necessary information and instructions to the Contractor (2).
- Dates for commencement and completion are to be entered in the Conditions, and at the Adviser's discretion he or she may revise the completion date where completion is delayed due to circumstances beyond the Contractor's control.
- In the event of failure to complete to time, liquidated damages are payable at the figure entered in the contract on a per week or part thereof basis (4).
- Practical Completion is to be confirmed in writing by the Adviser, and this term is defined in a 'note' in the contract (4·3).
- There is a defects period of six months and at the end the Adviser is to list defects 'he considers due' for making good by the Contractor (4·4).
- Assignment is not permitted without written consent of the other party, and the Contractor is not allowed to sub-let any work without the written consent of the Adviser (5).
- The contract is fixed price, except for any changes in statutory payments (7·3).
- The Adviser may order variations, and the order must also stipulate the basis for valuation (7·1).
- Payment to the Contractor will be as agreed and entered in the contract, that is either in a single main payment (6·2) or by interim payments made from time to time at the discretion of the Adviser (6·3).
- Payment is to be made within a stated period (entered in the contract) of certificates issued by the Adviser. A retention of 5 per cent (or other as entered) will be held by the Employer until the Final Certificate (6·4 and 6·5). Retention money is held in trust.

ASI Mini Forms

- The Final Certificate would appear to be conclusive evidence that work, materials and goods are in accordance with the contract (6·6).
- The Contractor must comply with all legal and statutory requirements, and pay all fees and charges (1·2). Presumably these will be added to the Contract Sum.
- The Contractor indemnifies the Employer and carries insurance cover for damage to property (other than the Works?) personal injury or death. Amount of cover to be entered in the contract. The Contractor is responsible for damage, loss or injury to the Works, unfixed materials etc. and must maintain a Contractor's all risks policy in joint names until the end of the Defects Liability Period (8·2).
- The contract provides for determination of the employment of the Contractor, for various listed reasons on the part of the Employer or the Contractor (9).
- Claims or disputes or other queries are to be referred in writing to the Adviser initially for a decision. If not resolved then they may be referred to adjudication (10·1).
- The adjudicator may be a person agreed between the parties, or nominated by the ASI President. The procedures for adjudication are succinctly stated in the contract, and the decision of the adjudicator is to be regarded as final and binding upon the parties (10).

2 ASI Mini Form of Contract (Home Improvement Agencies)

The wording of the Conditions in this version of the form are virtually identical with those of the blue form for General Use, with the following exceptions:

- The Letter of Invitation in the yellow form refers to the appointment of an 'Agency', and the letter is signed by the Agency, and not the 'Adviser/Employer' as is the case for the blue form. The Letter of Acceptance in each case in signed by the Employer.
- Clause 2·3 in the blue form refers to the appointment of a replacement Adviser. There is no such provision in the yellow form.
- Clause 7·3 in the blue form refers to statutory increases being acceptable in an otherwise fixed price contract. There is no such fluctuations provision in the yellow form, thus making it literally a fixed price contract.
- Clause 9.2 headed 'Determination by Contractor', refers in 9·2 ii) to delays of '4 weeks or longer, except by prior agreement' in the blue form. In the yellow form 9·2 ii) refers to delays of '4 weeks or longer at the Employer's insistence, except by previous agreement'.

ASI Mini Forms

This contract?

If considering using an ASI Mini Form remember that:

It is intended only for very small domestic work, hence the payment procedures and adjudication provisions are simpler than would otherwise be required to comply with Part II of the Housing Grants, Construction and Regeneration Act 1996. In both versions of the form the Employer must appoint an Adviser.

The yellow form (Home Improvement Agencies) is suitable for use in England and Wales. The blue form (General Use) is suitable for use in England and Wales, and an Amendment Sheet is available for use in Scotland.

The provisions are probably adequate for most work of this nature, but perhaps those concerning insurance and determination are a little light. The form is plainly worded.

The dedicated Letter of Invitation, Form of Tender and Letter of Acceptance should be used, because the last two become contract documents. When completing the documents, entries are also to be made in the text of the Conditions.

If acting as contract administrator ('the Adviser' is also the person who prepares the drawings) note that the procedural rules are relatively simple and the Adviser is given considerable discretion and authority. Care is needed before issuing the Final Certificate which would appear to be conclusive.

The ASI deserve credit for being the first to produce such forms, which also provide for a consultant to act as contract administrator (not necessarily an ASI member, nor indeed an architect). The forms are more conventional than the more consumer friendly later JCT forms.

Related matters

Documents

ASI Mini Form of Contract (General Use)
ASI Mini Form of Contract (Home Improvement Agencies)

Traditional procurement 9

Measurement forms

The Joint Contracts Tribunal Ltd
Standard Form of Building Contract 1998 Edition
With Approximate Quantities

The Institution of Civil Engineers
ICE Conditions of Contract Seventh Edition (1999)

The Institution of Civil Engineers
ICE Conditions of Contract for Minor Works Third Edition (2001)

The Joint Contracts Tribunal Ltd
Standard Form of Measured Term Contract 1998 Edition

For the purposes of this chapter, contracts which are not based on a lump sum figure and which provide for a substantial amount of remeasurement have been included as measurement contracts. It is acknowledged that most lump sum contracts accept remeasurement to some degree, and that SMM Rules provide for 'an approximate quantity' (which of course is not the same thing as 'approximate quantities'.) With measurement contracts the nature and extent of the work is broadly known before work is started, and this is not generally the case with 'cost plus' forms. These are dealt with separately in Chapter 10 below.

JCT 98 With Approximate Quantities

The Joint Contracts Tribunal Ltd

Standard Form of Building Contract 1998 Edition With Approximate Quantities

Background

There is often insufficient time (even when the scope of the work is reasonably definable and measurable right from the outset) to complete the drawings and Specification in sufficient detail to allow the quantity surveyor to fully measure, work up and collate measurements in order to prepare bills of quantities on behalf of the Employer for the purpose of obtaining lump sum tenders. In such circumstances the JCT Standard Form of Building Contract With Approximate Quantities may be appropriate.

This version of the standard JCT form was first produced in 1979 in response to pressure from property developers. They wanted a traditional method of building procurement, with as early a start on site as possible. It was thought that this could still be achieved if work could be described in accordance with SMM Rules, even though the quantity of work could not be accurately determined.

Nature

Developers often considered that the need for early completion outweighed the need for commitment to a firm price before work could start on site. Subsequent events suggested that this need may have been overstated and that the implications of not having a firm financial commitment had not been fully appreciated. This would seem to be borne out by the fact that initially fewer copies of the Approximate Quantities edition were sold than expected.

Use

The headnote to the Approximate Quantities edition states that the form is for use 'where the Works have been substantially designed but not completely detailed so that the quantities shown in the Bills are approximate and subject to remeasurement'.

Reference is made to bills of quantities setting out 'a reasonably accurate forecast of the quantity of work to be done'. The form should not be used where only certain sections of the Works are approximate. In such cases the Standard Form With Quantities should be used, and the relevant items might be marked 'Provisional' or, if under SMM7, could be the subject of 'an approximate quantity'.

JCT 98 With Approximate Quantities

Brief synopsis

- As might be expected the text of the With Approximate Quantities version follows very closely that of JCT98. The wording of the First and Second Recitals differs, and Article 2 refers to an Ascertained Final Sum instead of a Contract Sum.
- The With Approximate Quantities version differs from JCT98 in clauses 13 and 14 and the valuation of variations does not include for Price Statements or 13A Quotations.
- In clause 25·4·13 'work whose quantity was not reasonably accurately forecast in the Contract Bills' is included as an additional relevant event justifying an extension to the contract period.
- The list of matters for which loss and expense may be payable under clause 26·2·8 refers to work 'whose quantity was not reasonably accurately forecast in the Contract Bills'.
- Interim Certificates cover 'all work measured and valued by the quantity surveyor'. Valuations are always required under the Approximate Quantities form, and are to the usual valuation procedures. All documents and computations are to be sent to the quantity surveyor by the Contractor for the Final Sum.
- There is no provision for contracts with bills of approximate quantities to be let on a fixed price basis, and consequently clause 38 is not used. Fluctuation provisions will always apply.

There is a contracting out of third party rights under the Contracts (Rights of Third Parties) Act 1999 (1·12).

JCT 98 With Approximate Quantities

This contract?

If considering using JCT98 With Approximate Quantities remember that:

It is intended for use in substantial contracts where it is not possible to prepare full quantities at tender stage. It permits a certain amount of fast tracking, but the design needs to be well developed and main detail worked out before approximate quantities are taken off. This allows for some design development in parallel with construction work, but if the approximate quantities in the bills are not reasonably accurate there can be penalties in both cost and time. As with the other JCT98 forms there are different versions for private and local authority use. The Employer is required to appoint a Contract Administrator and a quantity surveyor.

If used for work in Northern Ireland an Adaptation Schedule should be incorporated, while for work in Scotland the Scottish Building Contract version of the form should be used.

The form has the same basic structure as the other JCT98 versions, using the same language and terminology. The main difference is the use of an ascertained Final Sum instead of a contract lump sum. The procedures relating to valuation are simpler than those in the other versions, but otherwise the payment provisions are similar.

It can include for partial possession, deferment of possession, sectional completion and performance specified work. Contractor's designed portion is not available for this version.

It allows for sub-contractors to be chosen by the Contractor from a list of not less than three names, and for nominated sub-contractors and suppliers. The dedicated NSC documents must be used with nominated sub-contractors.

When completing the form decisions are required relating to matters including deferment of possession; bonds (whether in lieu of retention, advance payment or 'listed items'); insurance of the Works; Joint Fire Code; liquidated damages; advance payment; fluctuations; and EDI. Care is needed to ensure that the relevant Supplements are used, and option clauses selected.

If acting as Contract Administrator note that some of the procedural rules are detailed and likely to prove time-consuming, particularly where there are nominated sub-contractors and suppliers.

Amendments are issued by the JCT from time to time. The form is available on disk. The RIBA publishes contract administration forms for JCT98, many (though not all) of which can be used with this Approximate Quantities version.

This form is probably the least risky remeasurement option for the Employer, and allows a measure of control not found in cost plus contracts. However, successful use depends very much on how full and accurate the approximate quantities are in the first place.

9 Traditional procurement: measurement forms

JCT 98 With Approximate Quantities

Related matters

Documents

Standard Form of Building Contract 1998 Edition
Private With Approximate Quantities
Local Authorities With Approximate Quantities
Amendment 1: 1999 (Construction Industry Scheme)
Amendment 2: 2000 (sundry amendments)
Amendment 3: 2001 (terrorism, Joint Fire Code, CIS, SMM)
Amendment 4: 2002 (extension of time, loss and expense, advance payment)

Nominated sub-contractor forms:
NSC/T 98: Part 1 (tender invitation, architect to sub-contractor)
NSC/T 98: Part 2 (tender)
NSC/T 98: Part 3 (particular conditions of tender)
NSC/A 98 (agreement, contractor and sub-contractor)
NSC/N 98 (nomination from architect to contractor)
NSC/W 98 (warranty agreement, employer and sub-contractor)
NSC/C 98 (conditions of sub-contract)
(all these documents have to be used if nominating sub-contractors)

Nominated supplier forms:
TNS/1Tender
TNS/2Warranty

JCT Domestic Contracts: (no requirement in contract that these particular forms be used)
Sub-contract Agreement DSC/A/SC 2002 Edition
Sub-contract Conditions DSC/C/SC 2002 Edition

Supplements:
Fluctuations: Private (includes for approximate quantities)
Fluctuations: Local Authorities (includes for approximate quantities)
Sectional Completion Supplement (with quantities/approximate quantities)

References

Earlier JCT Practice Notes (relevant to JCT98 but written with JCT80 in mind):
Practice Note No 23: Contract Sum Analysis (1987)
Practice Note No 24: Insolvency of Main Contractor (1992)
Practice Note No 25: Performance Specified Work (1993)
Practice Note No 27: Application of the CDM Regulations (1995)
Practice Note No 28: Mediation (1995)
Series 2 JCT Practice Notes (yellow covers):
Practice Note 1: Construction Industry Scheme
Practice Note 2: Adjudication (includes text of agreements)
Practice Note 3: Insurance, Terrorism Cover
Practice Note 4: Partnering (includes text of non-binding charter)
Practice Note 5: Deciding on the Appropriate JCT Form of Contract
Practice Note 6: Main Contract Tendering (includes model forms)

Commentaries (relating to JCT98)

David Chappell
Parris's Standard Form of Building Contract
3rd edn, Blackwell Science (2002)

Sarah Lupton
Guide to JCT98
RIBA Publications (1999)

JCT in collaboration with Building Design Partnership
Guide to the Use of Performance Specifications
RIBA Publications (2001)

RIBA Publications
Architect's Guide to the Contract Administration Forms JCT98
RIBA Publications (2000)

ICE Conditions

ICE Conditions

ICE Conditions of Contract Measurement Version Seventh Edition (September 1999)

Background

The sub-title of the document is Conditions of Contract and Forms of Tender Agreement and Bond for use in connection with Works of Civil Engineering Construction: Measurement Version.

This is a form which is clearly suited to the scale and relatively imprecise nature of civil engineering operations, which make remeasurement inevitable. The Conditions show that the form is not really intended for building work and that it is not appropriate for lump sum contracts.

This ICE form is sponsored by the Institution of Civil Engineers, the Association of Consulting Engineers and the Civil Engineering Contractors Association. It is kept under review by a standing committee and revised 'when such action seems warranted'.

It is of course part of a range of forms quite distinct from the New Engineering Documents which, although also from the Institution of Civil Engineers, are the responsibility of its New Engineering Contract Panel. These forms continue to be produced in parallel, thereby providing for the needs of the modernisers, as well as the more traditional users.

Nature

The document is over 70 pages long and comprises the Conditions of contract; a Form of Tender with an Appendix in two parts, the first of which is to be completed prior to tendering, and the second of which is to be completed by the Contractor; and a Form of Agreement. The latter identifies the documents which form part of the contract and may be drawings, the Specification, and the priced bill of quantities, in addition to the Conditions.

Also included is the ICE Form of Default Bond, and the Contract Price Fluctuations for Civil Engineering Work and Structural Steelwork.

The Conditions are contained in 72 clauses, without any apparently logical sequence or structure, but with historic precedence. However, reference is made relatively easy for the uninitiated because there is a Table of Contents and a particularly detailed Index.

9 Traditional procurement: measurement forms

ICE Conditions

Use

The Form of Agreement identifies the contract documents, and entries in the Appendix to the Tender will show detail such as commencement date, contract period, completion by sections, liquidated damages, payment and retention, insurance requirements etc.

The key person for contract administration is 'the Engineer', who must be named, and in the event of his or her being unable to act the Employer is obliged to nominate a successor.

Synopsis

1 Intentions

- The Contractor undertakes to construct and complete the Works as specified in or inferred from the contract (8).
- The Contractor is responsible for all site operations and methods, but not the design of permanent works (unless expressly provided in the contract), nor for the design of temporary and permanent works designed by the Engineer. Any design responsibility of the Contractor is limited to using reasonable skill, care and diligence (8).
- The Contractor is to work in strict accordance with the contract, and to the satisfaction of the Engineer. Mode, manner and speed are to be acceptable to the Engineer (13).
- The definition of 'the Works' includes both temporary and permanent work. The 'Contract' means the Tender (which has an important Appendix) and written acceptance of it, drawings, priced bill of quantities, Specification, and the Contract Agreement (1). The documents are to be taken as being mutually explanatory (5).
- The quantities in the bill are estimated only, and remeasurement will establish the actual quantities (55).
- The Contractor is expected to inspect and examine the site and its surroundings before tendering. The rates and prices he quotes will be in the bill of quantities, and only if matters which could not have reasonably been foreseen are encountered is additional payment possible (11 and 12).
- The contract allows for further necessary drawings, Specifications and instructions to be supplied by the Engineer from time to time (7).
- The Contractor is required to submit a Programme for approval by the Engineer, and to revise it within 21 days if the Engineer instructs (14).
- All materials and workmanship shall be as described in the contract, and as instructed by the Engineer. There is provision for samples and testing (36).
- Facilities must be afforded by the Contractor for work not in the contract but ancillary

to the Works undertaken by other contractors employed by the Employer (31).

- The duties and authority of the Engineer, the Engineer's Representative, and named assistants, are all clearly defined (2).

2 Time

- The contract may prescribe that possession of the site is to be given to the Contractor in portions (42).
- The Contractor is to commence as soon as possible after the Works Commencement Date as stated in the Appendix to the Form of Tender, or within 28 days after the contract is entered into, or other agreed date. The Contractor is then to proceed with expedition (41).
- The time for completion is to be stated in the Appendix to the Form of Tender. Completion by sections may be required (43).
- Interim extensions of time may be awarded. The Contractor must inform the Engineer within 28 days after the cause of the delay, and supply detailed particulars. The Engineer must respond to all claims upon receipt of particulars, and may award an extension in the absence of a claim. The overall extension of time awarded is subject to final review after completion (44).
- Where the Engineer thinks that progress is too slow and no extension of time is possible, written notice can be given to the Contractor to expedite matters. The Contractor will not be entitled to additional payment for taking such steps as may be necessary (46).
- The Employer will be entitled to liquidated damages if the Contractor does not complete to time (47).
- The Engineer will issue a Certificate of Completion when the whole of the Works are substantially completed and have passed any final tests required by the contract (48).
- The outstanding work and defects period runs from the date of substantial completion, and is stated in the Appendix to the Tender as the Defects Correction Period. The Contractor is obliged to finish any work outstanding, to deal with repair, amendment, reconstruction, rectification and making good of defects either within the period or as soon as may be practicable. A Defects Correction Certificate is to be issued by the Engineer when he or she is satisfied, although this in no way relieves the Contractor of any liability (61).

3 Control

- The Contractor is prevented from assigning the contract or any benefit or interest under it without the written consent of the Employer (3).
- The Contractor is not permitted to sub-contract any part of the Works including their design without giving written notice to the Engineer (4).

ICE Conditions

- Sub-contractors may be nominated, although the Contractor is given the right of objection (59). There is a detailed set of provisions for dealing with nomination and, as is normal, the Employer is required to bear some of the risk where defaults occur. If any design obligation rests with the sub-contractor, then this must be stated in the contract and the sub-contract (58[3]).
- The Contractor is wholly responsible for the accurate setting out of the Works and, unless incorrect information was given by the Engineer, must rectify any error at his own cost (17).
- The Engineer is empowered to suspend any part of the Works (40) and to order variations, subject to their being of the type defined in the contract (51).
- The Contractor is obliged to give the Engineer a reasonable opportunity to inspect work before it is covered up (38).
- The Engineer is empowered to instruct the removal from the site of any materials not in accordance with the contract, and the proper re-execution of work (39).
- The Contractor is obliged to provide necessary superintendence, and a competent agent or representative must be constantly on the Works (15).

4 Money

- Because this is not a lump sum contract, there is reference to a 'Tender Total' (i.e. total of the priced bill of quantities or agreed estimated total) and a 'Contract Price' (i.e. the figure finally ascertained) (1[1]).
- VAT will not have been allowed for in the Tender, and will be additional to the Tender Total (70[1]).
- Provisional sums and Prime Cost items are both defined (1[1]), and work or goods which are the subject of provisional sums and Prime Cost items may be ordered by the Engineer (58).
- Variations may be ordered by the Engineer (51). Wherever possible quotations should be agreed in advance of the order. Otherwise variations are valued at the rates in the contract if applicable, or otherwise at rates fixed by the Engineer, as the Engineer considers reasonable (52). Daywork is to be used as a basis if the Engineer thinks it necessary or desirable, and the rates will be as set out in the daywork schedule included in the contract (56[4]).
- Payment is on the basis of a monthly statement submitted to the Engineer (60[1]). The Engineer must issue a certificate within 25 days of delivery of a statement. Amounts for nominated sub-contractors are listed separately. The certificate must show the amount due and the basis of calculation. If the Employer intends to withhold payment, he must notify the Contractor not less than one day before the final date for payment, and state the grounds for any deduction. Final date for payment by the

ICE Conditions

Employer is 28 days after delivery of the Contractor's monthly statement to the Engineer (a quite tight timescale). If the Engineer fails to certify, or the Employer fails to make proper payment to time, then the Contractor is entitled to interest on the overdue amount (60[7]).

- Rate of retention, and the limit of retention will be as entered in the Appendix Part 1 (60[5]).
- Half the retention is released within 14 days of the issue of the Certificate of Substantial Completion, and the other half within 14 days of the end of the Defects Correction Period (60[6]).
- Not later than three months after the date of the Defects Correction Certificate, the Contractor is to give the Engineer a final account and supporting documents. The Engineer then has up to three months to issue the Final Certificate which states the amount due. Payment is required within 28 days of certification (60[4]).

5 Statutory obligations

- The Contractor is obliged to give all notices and pay all fees required by legislation. These might relate to temporary works. If the Engineer certifies it, then the Contractor will receive repayment of all sums involved (26).
- Whilst the Contractor indemnifies the Employer against the consequences of any breach of statutory obligations, this will not apply if it arises because of compliance with instructions given by the Engineer. There is no express obligation on the Contractor to check whether documents or instructions by the Engineer conform to statutory requirements, but if they are found not to, the Engineer must issue instructions (26).
- The CDM Regulations 1994 place statutory obligations on both Employer and Contractor, particularly in respect of the Health and Safety Plan and the Health and Safety File. The Employer is obliged to appoint a Planning Supervisor and a Principal Contractor. Through incorporation of these provisions into the contract, the statutory duties also become contractual obligations (71).
- The New Roads and Street Works Act 1991 is given detailed attention (27). There is also particular reference to legislation on damage to highways (30) and unnecessary interference to roads and footpaths (29).

6 Insurance

- The Contractor is to take full responsibility for the care of the Works and materials for incorporation, from the date of commencement until the Engineer has issued the Certificate of Substantial Completion. The risk covers damage from any cause whatsoever, except where the damage is due to the Employer, or is a defined Excepted Risk, or is due to faulty design of the permanent works (20).

ICE Conditions

- The Contractor is to insure in joint names, against the risk of damage to the Works for the full replacement cost plus an additional 10 per cent. The policies are subject to approval by the Employer before Works Commencement Date, and it may be prudent for the Employer to state the kind of cover required at tender stage (21).
- The Contractor indemnifies the Employer against all losses and claims arising from death or injury to any person, and damage to property other than the Works. There are certain stated exceptions (22).
- The Contractor is required to be insured against many of the risks in respect of which the Employer has been indemnified (23).
- The Contractor's obligations concerning the safety of workmen is reinforced, and the Employer is indemnified in respect of any claims (24).

7 Termination

- The Employer is given the right to terminate the employment of the Contractor in the event of the Contractor's insolvency, or because of specified defaults by the Contractor. The Employer is entitled to take possession of the site, have the work completed, and postpone any settlement of accounts until the Defects Correction Period has expired (65).
- The Contractor is given the right to terminate his own employment for specified defaults by the Employer, or if the Employer becomes insolvent (64).
- Termination of nominated sub-contracts is dealt with separately (59).

8 Miscellaneous

- Definitions are included (1).
- The Engineer can require the removal of Contractor's employees (16).
- There is provision in the event of discovery of fossils, antiquities, things of archaeological interest, etc. (32).
- The Engineer is to have access to the Works, the site and workshops (37).
- Matters relating to possession of site and site access, beyond that prescribed in the contract, are largely the Contractor's responsibility (42), although extensions of time and additional costs may be applicable in the event of failure to give adequate possession.
- No equipment owned by the Contractor is to be removed from the site without written consent of the Engineer (54).
- The rights and obligations of the parties on the outbreak of war are fully stated (63).

There is a provision contracting out of the Contracts (Rights of Third Parties) Act 1999 (3[2]).

ICE Conditions

- Special conditions can be properly incorporated and should be numbered consecutively as a continuation (72).

9 Disputes

- The terms of the contract require that matters of dissatisfaction must be referred to the Engineer in the first instance. The Engineer is required to state his or her decision in writing to both the Employer and the Contractor within one month of the reference (66).
- If a matter of dissatisfaction cannot be satisfactorily resolved by reference to the Engineer, then it becomes a dispute. The parties might, by agreement, seek consideration of the dispute under the ICE Conciliation Procedure (1999) (66(5)).
- Since the Housing Grants, Construction and Regeneration Act 1996 (Part II) came into force, parties to the contract have the right of reference to adjudication. With the ICE Conditions of Contract a notice of adjudication is to be given, and adjudication is to be conducted in accordance with the ICE Adjudication Procedure (1997) (66(6)).
- All disputes may be finally determined by reference to arbitration. The party seeking arbitration must serve a notice to refer. Arbitration is to be under the Arbitration Act 1996, and conducted in accordance with either the ICE Arbitration Procedure (1997) or the Construction Industry Model Arbitration Rules.

ICE Conditions

This contract?

If considering using ICE Conditions of Contract remember that:

This is a contract primarily intended for use with major civil engineering works, on the basis of measurement. It is not readily suitable for building work, but architects may find themselves involved in some secondary building work to which these Conditions will apply.

The form is for use under the law of England and Wales, and may also be adapted for use in Northern Ireland and under Scots law. If the law of the contract is that of Scotland or Northern Ireland, then Amendment Reference ICE/Scot/Arb/April 2001 should be incorporated.

The Conditions are comprehensive, but the terminology, although familiar to civil engineers, might need to be studied carefully by others. There is flexibility over commencement and completion, nomination of sub-contractors, valuation of variations, and design by the Contractor.

Completing the form is confined to entries in those documents which follow the Conditions, in particular the Appendix to the Tender and the Form of Agreement.

In terms of contract administration the Engineer is given greater authority than is usually the case with contract administration under building contracts.

It is the traditional self-contained form without the need for supplements. Much used and with a distinguished pedigree. It may lack a logical structure and some legal commentators have criticised it on the grounds of being rather imprecise, but it seems to have worked very well in practice for a very long time.

Related matters

Documents

ICE Conditions of Contract: Measurement Version 7th Edition (September 1999)
ICE Conciliation Procedure (1999)
ICE Arbitration Procedure (1997)
ICE Adjudication Procedure (1997)

Notes

Guidance Notes to the ICE Conditions of Contract 7th Edition

ICE Minor Works

ICE Conditions of Contract for Minor Works Third Edition (2001)

Background

The sub-title to this form is Conditions of Contract, Agreement and Contract Schedule for use in connection with Minor Works of Civil Engineering Construction. The sponsoring authorities are the Institution of Civil Engineers, the Association of Consulting Engineers, and the Federation of Civil Engineering Contractors.

Nature

This is a document intended for use where the risks are small, the contract period six months or less, there are no nominations, design of the Works is essentially complete, the Contractor has no responsibility for design of the permanent works, and the contract value does not exceed £250,000. It is a form suitable for straightforward and simple jobs.

The form is 20 pages long, and this includes the Conditions, an Appendix, and a Form of Agreement. The Conditions are set out under 13 headings, and are commendably clear and an easy read. They are prefaced by a full Index, and helpful Guidance Notes, not part of the contract, are included with the form. This latest edition takes account of Part II of the Housing Grants, Construction and Regeneration Act 1996.

Use

The form is suitable for lump sum contracts, or with measurement contracts using a priced bill of quantities, valuation based on a Schedule of Rates or a Daywork Schedule, or as a cost plus form. An Appendix entry will show which is applicable.

The Employer is required to appoint an Engineer to administer the terms of the contract. The form also provides for the appointment of a Planning Supervisor. A starting date is to be entered in the Appendix, together with a period for completion. Completion of the Works can be in phases.

Synopsis

1 Intentions

- This is essentially a contract for engineering rather than building work.
- The Employer must appoint an Engineer, who is to be named in the Appendix (2·1).
- The Engineer may appoint a named Resident Engineer, and delegate any powers except those dealing with disputes (2·2).

ICE Minor Works

- The contract means the Agreement (curiously qualified by the words 'if any'), the Conditions, the Appendix, and other items listed in the 'Contract Schedule' (a term not defined in the Conditions and the Contract Schedule which was included in the Second Edition of this form has now been omitted). Presumably this can be taken as referring to the drawings, Specification, priced bill of quantities, Schedules, and other documents now listed as part of the Form of Agreement (1·2).
- The Contractor undertakes to perform and complete the Works (3·1). He takes full responsibility for the care of the Works from commencement until 14 days after the issue of the Practical Completion Certificate (3·2).
- The Contractor is liable for design only where expressly stated in the contract, and in respect of temporary works other than those designed by the Engineer (3·7). The standard expected is that of reasonable skill, care and diligence.
- The Engineer must provide any necessary instructions, drawings or other information (3·6).
- Other persons engaged direct by the Employer are to be given reasonable facilities (3·9).

2 Time

- The starting date is entered in the Appendix; if not, then the Engineer must give written instructions within 28 days after acceptance of the tender (4·1).
- The contract period will be stated in the Appendix (4·2).
- If required, the Contractor must provide a programme within 14 days of the starting date (4·3).
- An extension of time may be granted where progress is delayed for reasons stated in the Conditions (4·4) which the Contractor has taken all reasonable steps to avoid.
- Practical completion occurs when the Works are fit to be taken into use or possession by the Employer. The Engineer is to certify Practical Completion, or must advise the Contractor in writing what remains to be done to achieve it. Partial possession is possible (4·5), as is completion by sections identified in the Appendix.
- Liquidated damages will be payable where the Contractor fails to complete by the completion date (4·6).
- A Defects Correction Period follows Practical Completion. The duration of this will be stated in the Appendix (5·1).
- Completion is to be certified by the Engineer after the defects have been made good and the Defects Correction Period has expired. Certification is in response to a request by the Contractor, and shows that the contractual obligations have been discharged to the Engineer's satisfaction (5·4).

ICE Minor Works

3 Control

- Instructions which the Engineer is empowered to issue are listed. These include variations, testing, suspension of work, removal of work not in accordance with the contract and exclusion of persons (2·3).
- Each of the parties is bound by every instruction or decision of the Engineer, unless it concerns a matter referred for dispute resolution (2·7).
- The Contractor cannot assign the contract or any rights under it without the written consent of the Employer (8·1). The Employer equally requires the Contractor's written consent.
- The Contractor may not sub-contract any part of the Works without the consent of the Engineer (8·2).
- There is no reference to nominated sub-contractors or suppliers, but the Guidance Notes suggest that approved sub-contractors and suppliers can be named and listed by the Engineer in the Specification.
- The Contractor accepts full responsibility for setting out and for the stability and safety of his operations and methods (3·5).
- The Contractor is to notify the Engineer of the person who is authorised to receive instructions (3·4).

4 Money

- More than one basis of payment can be used on any one contract. The Appendix requires deletion of methods not used. Options are for lump sum, measurement on priced bill, valuation on a Schedule of Rates, valuation on a Daywork Schedule and cost plus.
- VAT is not included in the tender figure.
- The Contractor may be entitled to additional payments due to unforeseeable adverse conditions (3·8) or because of delay or disruption to progress (6·1).
- The Contractor is to submit a monthly statement giving the value of work executed and materials or other items to be included (7·2).
- The Engineer is to issue an Interim Certificate, within 25 days of delivery of the monthly statement (7·3). The Appendix may include a minimum figure for any Interim Certificate. Every Certificate shall show the amount due and the basis on which it was calculated. Payment by the Employer becomes due on certification, with a final date for payment 28 days after delivery of the monthly statement. If the Employer intends to withhold payment, he must notify the Contractor not less than one day before the final date for payment, and state the grounds for any deduction (7·10).

ICE Minor Works

- Within 28 days after the Engineer has certified completion, following making good of any outstanding work at the expiry of the Defects Correction Period, the Contractor shall submit a final account. Within 42 days of receipt of this the Engineer should issue the Final Certificate. The amount payable is due upon certification, and the final date for payment is 14 days later (7·6).
- Half the retention is to be released within 14 days of the issue of the Practical Completion Certificate, the remainder to be released on certification of final completion (7·4 and 7·5).
- Interest may be added to overdue payments (7·8).

5 Statutory obligations

- The Contractor is required to comply with all relevant statutory obligations, give all notices required, and pay all fees and charges (9·1).
- Responsibility for any consent, approval, licence or permission for permanent works rests with the Employer. This would include planning consent and also presumably matters of land law including party walls consents (9·2).
- The Contractor's liability does not extend to breach where this has occurred because instructions from the Engineer have been followed (9·3).
- The CDM Regulations 1994 place statutory obligations on both Employer and Contractor. Unless stated otherwise, the Employer nominates the Engineer to act as Planning Supervisor and the Contractor as Principal Contractor. There are requirements under the Regulations concerning the Health and Safety Plan and the Health and Safety File, and these obligations become contractual as well (13).

6 Insurance

- If insurance of the Works is to be the responsibility of the Contractor an Appendix entry is required. The Contractor is then obliged to take out joint names insurance in respect of temporary and permanent works. Unfixed materials and construction plant are included. Cover is for full value against all loss or damage, bar Excepted Risks as defined in the contract Conditions (10·1).
- The Contractor gives the Employer an indemnity against loss and claims for injury or damage to persons and property. This liability will be reduced proportionately to the extent that the Employer or those for whom he is legally responsible contributed to the cause (10·2).
- There are further exceptions to the Contractor's liability for matters beyond his control and in respect of which he does not indemnify the Employer (10·4).
- Insurance required under the contract is subject to approval by the Employer (10·6).

ICE Minor Works

7 Termination

- There is no express provision for determining the Contractor's employment. The parties must rely on common law. The matter of what breach might constitute a repudiation of contract obligations is left unexplored.

8 Miscellaneous

- Definitions are included as part of the Conditions of contract (1).
- There is a contracting out of rights of third parties in accordance with the Contracts (Rights of Third Parties) Act 1999 (8·1).

9 Disputes

- The terms of the contract require that matters of dissatisfaction must be referred to the Engineer in the first instance. The Engineer is required to state his or her decision in writing to both the Employer and the Contractor within one month of the reference (Addendum clause A·2).
- If a matter of dissatisfaction cannot be satisfactorily resolved by reference to the Engineer, then it becomes a dispute. The parties might, by agreement, seek consideration of the dispute under the ICE Conciliation Procedure (1999) (Addendum clause A·5).
- Notice of referral to adjudication is open to either party, and adjudication is to be conducted in accordance with the ICE Adjudication Procedure (1997) (Addendum clause A·6).
- Disputes may be finally determined by reference to arbitration. The party seeking arbitration must serve a notice to refer. Arbitration is to be under the Arbitration Act 1996, and conducted in accordance with either the ICE Arbitration Procedure (1997) or the Construction Industry Model Arbitration Rules.

ICE Minor Works

This contract?

If considering using the ICE Minor Works form remember that:

It is intended primarily for minor engineering works, short duration, not exceeding £250,000. It is for lump sum with or without a bill of quantities, or for measurement contracts. The Employer is required to appoint an Engineer to be Contract Administrator. The contract is for use under the law of England and Wales, and is suitable for use in Northern Ireland or under Scots law as provided for in clause 12.

It is a truly short but balanced form which allows for phased completion. The Conditions are succinctly and clearly worded. Sub-contractors may be listed in the Specification.

Completing the form is straightforward, with entries required in the Agreement and the Appendix.

Contract administration should not be demanding as the Engineer is given considerable authority, and the procedures are straightforward.

A pleasing and neat form for low risk projects.

Related matters

Documents

ICE Conditions of Contract for Minor Works Third Edition (2001) (Reprinted 1998)

Notes

ICE Conditions of Contract for Minor Works Third Edition: Guidance Notes

JCT Measured Term

The Joint Contracts Tribunal Ltd

Standard Form of Measured Term Contract 1998 Edition

Background

This standard Measured Term Contract was introduced in 1989 for use by Employers in both public and private sectors who have building stock in need of planned regular maintenance and minor improvement work. It is obviously tiresome and wasteful having to enter into separate contracts for each small job, and in some cases a Contractor might be needed to undertake repair work at short notice. In all such circumstances it is usually preferable to deal with one Contractor appointed to handle all such work under one contract for a specific period, on terms previously agreed.

Competitive tenders may be invited on the basis of rates taking into account the nature of the intended works, the geographical area to be covered, and the duration of the contract. Measured term contracts have long been used by corporate and commercial client bodies with building stock, but this JCT contract is the first standard form specifically for such work.

Nature

The total document is 50 pages long, and contains a Contents table, Articles of Agreement with two Recitals and seven Articles, the Conditions, Supplementary Condition A, and an Appendix. The typographical presentation is commendably clear, and the Conditions are set out under nine section headings.

The First Recital refers to the nature of the intended work as being 'maintenance and minor'. It also requires the 'Contract Area' to be defined. The Employer is required to appoint a Contract Administrator, a person who has special significance under this contract because each separate job within the period of the contract must be initiated by an order issued by the Contract Administrator. The First Recital also refers to CDM Regulations compliance in respect of each separate order.

The Second Recital embodies the offer by the Contractor to carry out work under the contract to agreed terms of payment set out in the Appendix, and acceptance of this offer by the Employer.

The Articles incorporate the Conditions, and the information entered in the Appendix. Article 3 identifies the name of the person or firm acting in the capacity of Contract Administrator, and requires that any replacement needed must be nominated within 14 days. Article 4 identifies the Planning Supervisor and Article 5 the Principal Contractor. Entries in the Appendix are particularly important with this type of long term arrangement.

JCT Measured Term

Details should be given of the types of work for which orders may be issued, a list of properties in the Contract Area which may be included within the contract, and an indication of the estimated value of work which may be anticipated (although not guaranteed). The duration of the contract period, which will normally be for a minimum of one year, should also be entered.

The terms for measurement, valuation and payment are under items 6 to 10 of the Appendix. Item 11 refers to the Contractor's safety policy, which is to be attached to the form and thereby becomes a contractual obligation. This will relate to work carried out under the contract as a whole, whilst Health and Safety Plan requirements will be specific to work carried out under separate orders.

By nature the Measured Term Contract differs from a lump sum contract in that there can be no precise amount of work established at the outset and no Contract Sum. It is an enabling document, which allows for the issue of specific orders over a given period, to be valued according to rates, prices and percentages as entered in the Appendix.

Use

After the identity of the parties and the date of the Agreement, the Recitals require an entry which states the 'Contract Area'. There is no reference to drawings or documents other than the accepted terms of payment. The Contract Administrator must be named under Article 3. The term 'Architect' is not used at all in the contract.

The form can be executed under hand and not as a deed, or as a deed.

The Appendix requires an entry stating the minimum and maximum value of any one order to be issued, together with an approximate total figure or figure per annum. A priority coding for orders may be introduced, which would signal the need for rapid response by the Contractor to specific orders such as emergency repairs within stipulated minimum periods.

Note that for payment it is necessary to identify whether payment is based on (a) the National Schedule of Rates, or (b) some other Schedule of Rates. The contract can be fixed price as far as rates are concerned, or subject to fluctuations.

Responsibility for measurement and valuation can rest with either the Contractor or the Contract Administrator. Appendix entries should indicate whether the Contractor is to undertake all measurement and valuation, or conversely, that the Contract Administrator is to be responsible. Alternatively it can be shown that only orders above a stated value will be the responsibility of the Contract Administrator for measurement and valuation.

With a contract of relatively long duration, it is desirable that there should be a break provision. The period of notice required to bring this about is 13 weeks unless a different period is entered in the Appendix.

JCT Measured Term

Synopsis

1 Intentions

- The definitions in this contract have a particular significance, because of their context (1·1).
- The Contractor is obliged to carry out work when issued with an order. The work is to be executed in a good and workmanlike manner, in accordance with the Contract Documents, empowered instructions, and any relevant Health and Safety Plan. The Contractor does not have exclusive right to all work within the Contract Area and the Employer can use other contractors or his own labour force if desired (1·2).
- The Employer has the right to supply direct any materials, plant or equipment needed for carrying out work, and the Contractor will have responsibilities concerning these (1·6 to 1·11).
- Whatever the source of the materials, they must be to the standards described in the Schedule of Rates (1·12).
- The printed Articles, Conditions and Appendix entries prevail in the event of any conflict with Schedules, Specification or drawings (1·14).
- A Programme must be provided by the Contractor where the Contract Administrator requests one (1·15).
- If the Employer replaces the Planning Supervisor or the Principal Contractor, the Contractor must be notified (1·16).

2 Time

- The duration of the contract and its starting date are entered in the Appendix (item 4). There may be a priority code imposed.
- The starting date and reasonable completion date for work included under an order shall be stated in each order (2·2).
- Where matters beyond the control of the Contractor cause delay, an extension to the time for completion for each order may be given by the Contract Administrator on a fair and reasonable basis (2·3).
- There is a Defects Liability Period of six months in respect of each order, starting from the Order Completion Date (2·4).

3 Control

- The bar to assignment without written consent extends to the contract, or any part share or interest (3·1). Sub-contracting in any order is restricted to the extent that previous consent in writing of the Contract Administrator is required (3·2).

JCT Measured Term

- The Contractor is obliged to employ a competent representative (3·3). (Note: there is no reference to his or her being constantly or otherwise upon the Works.)
- Access to the site, which can be a complicated issue in work of this nature, lays a considerable burden upon the Contract Administrator. The Contractor is obviously entitled to a degree of possession sufficient to enable him to carry out the work under any order. Where access is restricted, unproductive time may be charged on a daywork basis (3·4).
- Variations may be instructed by the Contract Administrator, and what constitutes a variation under this contract is defined (3·6).
- An order, although issued, may be cancelled in writing by the Contract Administrator. The Contractor will be entitled to direct costs already incurred (3·8).
- The Contract Administrator has the power to order the exclusion of any person from the site (3·9).
- If the Contractor does not comply with written instructions from the Contract Administrator, then, after proper notice, other persons may be brought in by the Employer to give effect to the instruction at the Contractor's expense (3·10).

4 Money

- The work covered by each order is subject to measurement and valuation in accordance with the agreed Schedule of Rates (4·1). If the Schedule does not adequately allow for coverage, then fair rates or prices shall apply. In the event that agreement is still required, this shall be between the parties, or as a last resort it shall rest with the Contract Administrator, who must consult with the Contractor (4·3).
- If the Contractor's progress is interrupted because of an instruction issued by the Contract Administrator, then agreed lost time or other unproductive costs are valued on a daywork basis (4·7).
- Responsibility for measurement and valuation in the case of orders below an estimated value entered in the Appendix rests with the Contractor (4·8). Above this figure, unless there is anything to the contrary, responsibility will rest with the Contract Administrator (4·9).
- Most work carried out under an order will be subject to Part II of the Housing Grants, Construction and Regeneration Act 1996, and the contract provisions relating to payment take account of this.
- Progress payments at monthly intervals may be appropriate, depending upon the value of the order (4·11). In these cases the Contract Administrator should issue a certificate which states the amount estimated as due for payment, and also the basis on which this is calculated.

JCT Measured Term

- Where the Contract Administrator is responsible for measurement and valuation, a certificate for payment is to be issued within 56 days of the Order Completion Date (4·13). Where the Contractor is responsible for measurement and valuation, the account must be submitted to the Contract Administrator within 56 days of the Order Completion Date, and, if acceptable, a certificate is issued within 28 days of receipt of the account (4·14). In all cases the certificate must show the value of work completed under the order, together with the basis on which the value was calculated (4·13).
- The final date for payment by the Employer is 14 days from the date of issue of a certificate. Within five days of the date of the certificate the Employer shall given written notice to the Contractor of the payment to be made. If the Employer intends to deduct or withhold money, then the Contractor must be notified in writing of this intention no later than five days before the final date for payment.
- Failure to pay amounts properly due by the final date for payment will attract interest at the rate of 5 per cent over base rate current at the time payments by the Employer first became overdue.
- In the absence of any valid notices from the Employer, and without prejudice to any other rights and remedies, failure to pay by the date for final payment on any certificate can give the Contractor a right to suspend performance of all contractual obligations (i.e. presumably not confined to the particular order) (4·16).

5 Statutory obligations

- It is the Contractor's duty to comply with all statutory obligations and give all required notices (5·1). He is entitled to recover fees and charges not otherwise provided for (5·1). These duties arise in respect of any work undertaken in response to an order.
- The Contractor is to notify the Contract Administrator if he finds any conflict between statutory requirements and an order. The Contract Administrator must issue an instruction which will be a variation. The Contractor is then not liable for non-compliance resulting from the order or a subsequent variation (5·1).
- The respective obligations of the Employer and the Contractor, with regard to the CDM Regulations applying to an order, and in particular Health and Safety Plans and Health and Safety Files, become contractual as well as statutory duties (5·4).
- The Construction Industry Scheme (5·2) and VAT (5·3) are both covered under this section of the contract. Where the Employer is a 'contractor' for the purposes of the Scheme, Supplementary Condition A will be relevant.

6 Insurance

- The Contractor is to indemnify the Employer in respect of death or injury to persons or damage to property other than the Works, which arises from carrying out an order (6·1 and 6·2). This indemnity is to be backed by insurance, and an entry against item

JCT Measured Term

13 of the Appendix will state the minimum cover required.

- Insurance of existing structures and contents which might be affected by an order is the responsibility of the Employer. Cover will be for the full cost of reinstatement, repair or replacement after loss or damage due to one or more of the stated perils. Where the Employer does not wish to insure the existing structures, see model clause 6·5A (Annex D to the Measured Term Contract Guide).
- All risks insurance of work or supply instructed under orders is the responsibility of the Contractor. It will most likely be covered by an annual all risks policy, but this must be a joint names policy. Cover for each order is required up to the Order Completion Date (6·9).
- In the event that terrorism cover ceases to be available, the party responsible for insurance shall inform the other party, and the Employer must instruct the Contractor in writing which clause 6.14 option is to apply.

7 Termination

- The Employer is allowed to determine the employment of the Contractor for given reasons (7·1). Determination will affect the contract as a whole, although the default may arise only in respect of an order. The clause refers to work being 'significantly' suspended, disrupted or delayed, and minor or trivial instances would clearly be excluded by the words 'but not unreasonably or vexatiously' (7·1). A warning notice is required before the actual notice of determination by the Employer. In the case of insolvency, determination is automatic (7·2).
- The Contractor is allowed to determine his own employment for stated reasons, which include failure by the Employer to pay amounts properly due by the final date for payment (7·4).
- Common law rights are preserved whether determination is by Employer or Contractor, and the grounds stated in the contract are without prejudice to any other rights or remedies.
- The respective rights and duties of the parties concerning outstanding payments and any direct loss and/or damage arising from determination are set out in detail (7·6, 7·7).

8 Miscellaneous

- The contract includes break provisions, which allow for determination of the Contractor's employment by either party after six months. Such action could become necessary on a long-running contract because of changes to the Employer's building programme or due to a fluctuating workload. Thirteen weeks' notice (or lesser period if previously agreed) is required (8·1).
- Rights of third parties under the Contracts (Rights of Third Parties) Act 1999 are expressly excluded (1·20).

JCT Measured Term

9 Disputes

- Either party has a right to refer any dispute or difference arising under the contract to adjudication (Article 6 and clause 9A).
- Article 7B establishes arbitration as the agreed method for final determination of disputes (9B). Written notice of reference is required.
- The appointment of the arbitrator and conduct of the arbitration are clearly defined (9B). The parties agree that there may be appeal to the courts (9B·4).
- The JCT has adopted the 1998 JCT Edition of the Construction Industry Model Arbitration Rules, and the provisions of the Arbitration Act 1996 shall apply.
- Where it is stated that Article 7B applies, then the final determination of disputes is to be by legal proceedings and not by arbitration (9C).
- There is a footnote reference in clause 9A to 'mediation'. This means of resolving disputes may be particularly appropriate to the relatively low cost and short duration of work carried out under individual orders. It will not remove the statutory right to adjudication, nor a contractual agreement to arbitration or legal proceedings.

JCT Measured Term

This contract?

If considering using MTC98 remember that:

It is intended for maintenance and small works programmes of between one and three years' duration, and the approximate value of work to be carried out is given as a sum either per annum or relating to the contract period. There is only one version for both private and public sector use. The Employer is required to appoint a Contract Administrator. The form avoids the need for numerous separate contracts, but the appointed Contractor may only carry out each separate job when initiated by an order issued by the Contract Administrator.

The applicable law of the contract is the law of England, there the form may be used as printed for work in England and Wales. A footnote to clause 1·19 refers to use under other laws. There is an SBC Measured Term Contract (2001 Revision) for use under Scots law.

The form includes break provisions, commencement and completion of orders, arrangements for site access, and cancellation of an order. Orders are subject to a minimum and maximum value as entered in the Appendix to the form. Contract documents comprise the Articles, Conditions, Appendix and Schedule of Rates. There is one contract period and orders are given commencement and completion dates. All work is subject to measurement and valuation, and this can be made the responsibility of the Contractor or the Contract Administrator. Fluctuations provisions may apply or the contract may be fixed price.

When completing the form, Appendix entries are required on matters including list of properties; description of types of work; value of works; contract period; priority coding; payment provisions; and the break provision.

If acting as Contract Administrator, this responsibility will extend to all work carried out under the orders. This will include the issuing of orders, and may include issue of instructions, variations, certificates for progress payments, extension of time, and cancellation of an order. Note that a Contractor's safety policy for the contract is to be attached to the contract Appendix, in addition to the Health and Safety Plans which might become necessary for work carried out under each order to which the CDM Regulations apply. There is no standard format for orders, but the JCT Practice Note MTC/1 and Guide 1989 included a model Order Form and Completion Certificate in Annex E.

This is the only standard agreed form of contract published specifically for term contract working. The Conditions are clearly worded and section headed.

JCT Measured Term

Related matters

Documents

Standard Form of Measured Term Contract 1998 Edition
Amendment 1: 1999 (Construction Industry Scheme)
Amendment 2: 2000 (sundry amendments)
Amendment 3: 2001 (terrorism cover/CIS)

References

Practice Note MTC/1 and Guide
(includes excellent annexed material on Schedule of Rates, assessing tenders and a model Order)

Series 2 JCT Practice Notes (yellow covers):
Practice Note 1: Construction Industry Scheme
Practice Note 2: Adjudication (includes text of agreements)
Practice Note 3: Insurance, Terrorism Cover
Practice Note 5: Deciding on the Appropriate JCT Form of Contract

Traditional procurement

10

Cost plus forms

The Joint Contracts Tribunal Ltd

Standard Form of Prime Cost Contract 1998 Edition

For the Employer who must proceed on an indication only of the price, and agrees to pay the actual cost of labour, materials and plant, plus an amount for the Contractor's overheads and profit. Several standard forms may be used in a cost plus situation but there appears to be only one published for use exclusively under this arrangement.

JCT PCC98

The Joint Contracts Tribunal Ltd

Standard Form of Prime Cost Contract 1998 Edition

Background

Repairs and rebuilding during and immediately after the Second World War called for fast and sometimes even emergency action. Contracts based on the cost of labour and materials plus a percentage addition to cover the Contractor's overheads and profit were an obvious answer at the time. Shortly after the war, the Simon Report pointed out that fixing Contractor's profits as a percentage of the cost of the work was not in the public interest and unlikely to be acceptable to building owners in the future.

In 1967 therefore, the JCT published a Prime Cost form which incorporated a 'fixed fee', thus introducing an element of competition whilst still ensuring that Contractors should be able to recover the whole of their costs for labour and materials. Fixed fees would only be sustainable if the architect was precluded from issuing instructions which altered the 'nature or the scope of the Works'.

In 1981 the JCT set up a working party to revise the form as essentially a Prime Cost document, but with the contract fee either fixed or a percentage basis as desired. The draft produced in 1983 used a section headed format. An attempt was also made to resolve problems which arose where there were changes to the scope of the Works. In such cases the Contractor was required not to increase labour and materials more than was reasonably necessary to carry out the Works.

This revised Prime Cost Contract was eventually published in 1992, and is currently in a 1998 Edition.

Nature

This is a heavyweight document over 150 pages long, including an Appendix, with an Annex 1 to the Appendix relating to bonds, an Annex 2 to the Conditions relating to EDI provisions, Supplemental Provisions in respect of VAT, and Modifications necessary for Sectional Completion under Article 7. These Modifications are quite extensive, and clauses throughout the contract which are affected as the result of including Sectional Completion are identified by small letter 's' in the margin. This arrangement of printing Sectional Completion provisions in the body of the text, although convenient in many ways, does make for a rather bulky document and arduous cross-referencing.

The Articles of Agreement include seven Recitals and nine Articles, and the Agreement may be executed under hand not as a deed, or as a deed if required.

JCT PCC98

The contract Conditions are set out under nine section headings, helpfully indexed in a Contents page at the front of the form. A key feature essential with this kind of contracting is the provision of eight Schedules. These are very important components, on which a great deal of reliance is placed. The Schedules are as follows:

First Schedule: description of the nature and scope of the Works, and list of drawings (if any);

Second Schedule: definitions of the Prime Cost – relating to general items, labour, site staff, materials and goods, plant and services, sundry costs, sub-contract works etc.;

Third Schedule: Contract Fee – fixed fee or percentage fee;

Fourth Schedule: estimate of the Prime Cost of items of work;

Fifth Schedule: items of work to be executed by domestic sub-contractors listed in the Specification;

Sixth Schedule: items of work to be executed by nominated sub-contractors;

Seventh Schedule: materials and goods to be supplied by nominated suppliers;

Eighth Schedule: items of work to be executed by the Employer or others direct.

These schedules should be fully and carefully completed because the nature of the intended work is likely to mean that the contract will be let on minimal information. There may or may not be drawings to accompany the Specification, and the contents of the Schedules are therefore vital for both tendering and valuation purposes.

Use

In the Recitals, after stating in general terms the nature of the Works (also to be described as clearly as possible in the First Schedule), reference is made to a Specification and any drawings (which if used, are to be listed in the First Schedule). The Architect/Contract Administrator responsible for the preparation of these is the person named in Article 3. The quantity surveyor is the person named in Article 4.

The Contractor undertakes to carry out and complete the Works (as described in the First Schedule) together with other items as instructed by the Architect (although presumably the scope of the contract may not be materially altered).

The Employer undertakes to pay the Contractor the Prime Cost (hence the importance of the definitions in the Second Schedule) and the Contract Fee (Third Schedule). There is provision for the fee to be revised if alterations in the nature or scope of the Works justifies this.

One criticism of this kind of contracting is the lack of incentive for the Contractor to work efficiently. Protection for the Employer is afforded (C1·5·1) in that the Architect

JCT PCC98

is empowered to disallow costs where the Contractor does not carry out the work as economically as possible, or uses a greater number of operatives than is reasonably required. However, this might not be easy to implement in practice. As far as labour is concerned it could be very tricky for an Architect to say 'you have charged for six carpenters but four could have done that job'.

The Simon Report concluded that 'the control to be exercised in a cost reimbursement is a form of supervision, veto, and accounting'. It is doubtful whether this would be practicable today, as in any event it would probably call for constant monitoring.

Synopsis

1 Intentions

- The Contractor is obliged to carry out and complete the Works in all respects in accordance with the Contract Documents and empowered instructions of the Architect (1·5).
- Protection for the Employer against wasteful use of labour and materials is a provision (1·5).
- Quality and standards, so far as procurable, are to be as described in the Specification (3·10).
- The Contractor must be provided with such further drawings or details as are reasonably necessary to explain and amplify the Specification (1·6). There are limits to the use of the Specification and all drawings, and the confidentiality of rates must be respected (1·9).
- There is provision for correction of discrepancies in or between documents (1·10).

2 Time

- Dates for possession and completion are to be entered in the Appendix. There is provision for deferment of possession for up to six weeks. The Contractor is to proceed regularly and diligently and complete on or before the completion date.
- Notice of delay must be given by the Contractor in writing (2·5). The Architect is empowered to award an extension of time should completion be delayed beyond the completion date, provided the reason is one or more of the Relevant Events listed. The Architect is to make in writing a fair and reasonable extension within 12 weeks of receipt of the notice and particulars.
- Failure by the Contractor to complete within the contract period is to be certified by the Architect (2·2). Liquidated damages are then recoverable by the Employer.
- Practical Completion is to be certified by the Architect (2·8).
- There is provision for partial possession by the Employer (2·9).

JCT PCC98

- The Contractor is obliged to rectify defects (2·10) which are notified by the Architect within 14 days of the expiry of the Defects Liability Period.

3 Control

- The bar to assignment of the contract without written consent extends to both parties (3·1).
- The Contractor may not sub-contract without written consent of the Architect (3·17).
- Sub-contractors may be named and included in a list of not fewer than three names in or annexed to the Specification (3·18).
- There is also provision for nominating sub-contractors (8A·1). There are specified forms and procedures, (i.e. NSC/T (PCC); NSC/W (PCC); NSC/N (PCC); NSC/A (PCC) with Conditions incorporated by reference).
- Architect's instructions must be in writing (3·3). Alteration of the nature or scope of work is not empowered.
- The Contractor is required to have a competent person-in-charge on the site constantly (3·6). The Employer is entitled to appoint a Clerk of Works who can issue directions, although these must be confirmed as instructions by the Architect to have effect (3·9).
- There is provision for opening up, inspection and testing (3·11 and 3·12) and the familiar JCT Code of Practice relating to the fair and reasonable operation of this provision is included at the end of Section 3 of the form.
- There is provision for work to be carried out by others engaged direct by the Employer (3·13) and such items of work are best listed in the Eighth Schedule, and relevant information included in the Specification.

4 Money

- There is no Contract Sum, but Article 2 refers to a Prime Cost, Contract Fee and any direct loss and/or expense ascertained.
- The Contract Fee is to be stated in the Third Schedule, but may be adjusted (4·10).
- The Contractor may make written application for reimbursement of direct loss and expense suffered under specified headings or 'matters'. Common law rights are preserved (4·13).
- Interim valuations are to be made by the quantity surveyor, and the Contractor must provide the quantity surveyor with necessary details of expenditure to enable valuations to be made (4·4).
- Certificates are to be issued monthly and the first date is to be as entered in the Appendix.

JCT PCC98

- Certificates should include for work properly executed, the cost of materials and goods properly on site, an instalment of the fixed fee, and may include materials stored off-site where these are 'listed items', subject to reasonable proof of ownership and if requested provision of a bond (4·6).
- Interim Certificates must state what the amount relates to, and give the basis of calculation (4·2). Retention is applied to all Interim Certificates at 5 per cent or as otherwise agreed.
- The final date for payment on Interim Certificates is 14 days from the date of issue. Not later than five days after the issue of a certificate, the Employer should give written notice to the Contractor specifying the amount he proposes to pay. If he intends to withhold money, he must give the Contractor written notice not later than five days before the date for final payment. If the Employer fails to pay amounts properly due by the date for final payment, the Contractor has a right to suspend work (4·3).
- Issue of the Final Certificate is no later than two months from the end of the Defects Liability Period, or making good defects, or from receiving certain necessary information from the Contractor, whichever is latest (4·12).
- The Final Certificate must state the amounts due and the basis of calculation. Not later than five days after the issue of the certificate, the Employer is to give the Contractor written notice specifying the amount he proposes to pay. If he intends to withhold money, he must give the Contractor written notice not later than five days before the final date for payment (4·12). Final date for payment is 28 days from the date of the certificate (note not date of issue).

5 Statutory obligations

- It is the Contractor's duty to comply with all statutory obligations and give all required notices (5·1).
- The Contractor is to notify the Architect if he finds any conflict between statutory requirements and the documents. The Architect must issue an instruction and the Contractor is thereafter not liable to the Employer under the contract for any non-compliance with statutory requirements resulting from the instruction (5·2 and 5·5).
- The Contractor is empowered to carry out limited work for emergency compliance, and this will be treated as a variation and valued accordingly (5·4).
- Where the CDM Regulations apply, the Contractor is contractually obliged to comply with all the relevant CDM duties, particularly in respect of the Health and Safety Plan and in providing information for the Health and Safety File (5·20).

6 Insurance

- The Contractor indemnifies the Employer in respect of personal injury or death, unless due to the Employer's negligence (6·1).

JCT PCC98

- The Contractor indemnifies the Employer in respect of damage to property, provided that this is due to the Contractor's negligence (6·1).
- The Contractor is to maintain insurance to cover these indemnities, and the minimum cover required by contract is the sum entered in the Appendix (6·2).This does not necessarily limit the Contractor's liability.
- If instructed, the Contractor is to take out joint names insurance for the Employer against the risk of legal nuisance. There is a list of exceptions, and damage must not be attributable to any negligence by the Contractor (6·2).
- Insurance of 'the Works' is for all risks where new buildings are concerned, and a joint names policy may be taken out by the Contractor or the Employer as selected (6·3A or 6·3B).
- Insurance of existing structures and contents is for specified perils, and is to be taken out in joint names by the Employer (6·3C).
- In the event of terrorism cover ceasing to be available, the options open to the Employer are stated in clauses 6.3A, 6.3B and 6.3C, whichever is applicable.
- Where insurance is required against the Employer's loss of liquidated damages due to an extension of time following damages to the Works, then this should be entered in the Appendix (6·3D).
- An Appendix entry will show whether the Joint Fire Code of Practice on the Protection from Fire of Construction Sites is to apply. If so, both Contractor and Employer will need to respect it for in the event of non-compliance, the insurers can specify remedial measures which must be undertaken (6·3FC).

7 Termination

- The Employer is allowed to determine the employment of the Contractor by reasons of specified defaults (7·2). A warning notice may be issued by the Architect, but the notice of determination is a matter for the Employer. In case of insolvency of the Contractor, then depending on the circumstances determination might be automatic subject to possible reinstatement, or the Employer might enter into an agreement with the Contractor (a '7·5·2·1 agreement') for continuation or novation (7·5).
- Depending upon the circumstances, the Employer can have the right to make interim arrangements for certain work to be carried out during the holding period under this agreement. He may still elect to have the works completed by a different contractor, or abandon the idea of completing the Works altogether.
- The Contractor is allowed to determine his own employment for specified defaults by the Employer (7·9). The procedures must be followed meticulously. In the event of insolvency of the Employer, the Contractor may elect to determine his own employment.

JCT PCC98

- Either party can determine the employment of the Contractor for listed neutral causes (7·13).
- The respective rights and duties of the parties concerning payment, removal and completion are set out in detail in the applicable clauses.

8 Miscellaneous

- A full list of definitions, many of them specific to PCC98, is included (1·3).
- Special provisions apply to the keeping of records, measurements and accounts concerning the ascertainment of Prime Cost items (1·12).
- Section 8A of the form is given up to nominated sub-contractors, and Section 8B is given up to nominated suppliers.
- Access for the Architect is allowed at all reasonable times (3·8).
- Action necessary as a consequence of discovery of antiquities (3·16) is included.
- Third party rights under the Contracts (Rights of Third Parties) Act 1999 are excluded (1·25).

9 Disputes

- The Housing Grants, Construction and Regeneration Act 1996 (Part II) gives either party a statutory right to refer any difference or dispute arising out of the contract to adjudication. Article 8 of PCC98 provides for this.
- Procedures for adjudication are set out in detail in clause 9A.
- Arbitration is to be the method for final resolution of disputes (unless the Appendix statement has been deleted) and is detailed in clause 9B.
- Arbitration is to be conducted in accordance with the JCT 1998 Edition of the Construction Industry Model Arbitration Rules.
- Where arbitration has been deleted as the chosen method, then clause 9C will apply and the dispute will be determined by legal proceedings.

JCT PCC98

This contract?

If considering using PPC98 remember that:

It is intended for contracts where work must be started on site ahead of full documentation, or where the nature of the intended work is such that it is impossible to prepare full information. It is published in one version for either private or public sector clients. The Recitals call for a Specification and such drawings as are listed in the First Schedule, and an estimate of the Prime Cost of the items of work as shown in the Fourth Schedule. The Contract Fee charged by the Contractor may be either on a fixed or percentage basis as indicated in the Third Schedule. The Employer is required to appoint a Contract Administrator and a quantity surveyor.

The form is for use in England and Wales. The Scottish Building Contract Committee published its own SBC Prime Cost Contract 1999, but this is now apparently withdrawn.

PPC98 provides for partial possession, sectional completion (Modifications to facilitate this are included at the end of the contract form). There is no provision for design by the Contractor, nor for performance specified work.

Sub-contractors may be selected by the Contractor from a list of not less than three names. Sub-contractors and suppliers may also be nominated, in which case the use of dedicated documents is mandatory. The procedures follow closely those for JCT98, but the prescribed NSC documents carry the affix PCC.

When completing the form, entries are required relating to decisions on matters including deferment of possession; bonds (listed items); insurance of the Works; Joint Fire Code; liquidated damages; and EDI. The eight Schedules should also be checked for completeness of entries.

If acting as Contract Administrator it is essential to follow carefully the procedural rules, particularly where nominations are made.

Amendments are issued by the JCT from time to time. The form is available on disk. The RIBA does not publish contract administration forms for use with PCC98 but many of those for JCT98 can be adapted.

This is an attractively presented, self-contained but bulky document. The Conditions are section headed and sub-headed, but the procedures are rather complex for administration. It seems not to have sold in great numbers, not surprising perhaps as this is a procurement method which brings considerable risks for the Employer. However, it seems to have been used without serious problems, possibly due to careful selection of the Contractor, a realistic attitude on the part of the Employer and sound practice in contract administration.

JCT PCC98

Related matters

Documents

Standard Form of Prime Cost Contract 1998 Edition (PCC98)
Amendment 1: 1999 (Construction Industry Scheme)
Amendment 2: 2000 (sundry amendments)
Amendment 3: 2001 (terrorism cover/Joint Fire Code/CIS)
Amendment 4: 2002 (extension of time/loss and expense)

Nominated sub-contractor forms:
NSC/T 98 (PCC): Part 1 (tender invitation, architect to sub-contractor)
NSC/T 98 (PCC): Part 2 (tender)
NSC/T 98 (PCC): Part 3 (particular conditions of tender)
NSC/A 98 (PCC) (agreement, contractor and sub-contractor)
NSC/N 98 (PCC) (nomination from architect to contractor)
NSC/W 98 (PCC) (warranty agreement, employer and sub-contractor)
NSC/C 98 (PCC) (conditions of sub-contract)
(all these documents have to be used if nominating sub-contractors)

Nominated supplier forms:
TNS/1Tender
TNS/2Warranty
JCT Domestic Contracts: (no requirement in contract that these particular forms be used)
Sub-contract Agreement DSC/A/SC 2002 Edition
Sub-contract Conditions DSC/C/SC 2002 Edition

References

JCT Practice Note PCC/1 and Guide (for the 1992 Edition but largely relevant for 1998)

Earlier JCT Practice Notes (relevant to JCT98 but written with JCT80 in mind):
Practice Note No 24: Insolvency of Main Contractor (1992)
Practice Note No 27: Application of the CDM Regulations (1995)
Practice Note No 28: Mediation (1995)

Series 2 JCT Practice Notes (yellow covers):
Practice Note 1: Construction Industry Scheme
Practice Note 2: Adjudication (includes text of agreements)
Practice Note 3: Insurance, Terrorism Cover
Practice Note 5: Deciding on the Appropriate JCT Form of Contract

Commentaries

RIBA Publications
Architect's Guide to the Contract Administration Forms JCT98
RIBA Publications (2000)

(Although these are not forms specifically for PCC98, many of them, used with care, can be adapted for use with PCC98.)

Design and build procurement

11

Design and build forms

The Joint Contracts Tribunal Ltd
Standard Form of Building Contract With Contractor's Design 1998 Edition

The Joint Contracts Tribunal Ltd
Standard Form of Building Contract With Quantities 1998 Edition
Contractor's Designed Portion Supplement 1998

Institution of Civil Engineers
ICE Design and Construct Conditions of Contract Second Edition (2001)

The Stationery Office
GC/Works/1 Single Stage Design and Build (1998)
GC/Works/1 Two Stage Design and Build (1999)

Traditional procurement forms of contract do not, in the absence of anything to the contrary, provide for design by the Main Contractor. They are simply 'work and materials' contracts.

The wording used in some standard forms of contract expressly includes for a limited measure of design responsibility by the Main Contractor, sometimes by optional or supplemental provisions.

With one exception, this chapter is concerned only with forms of contract which have been drafted specifically for use with design and build procurement, and take full account of design responsibility by the Contractor.

JCT WCD98

The Joint Contracts Tribunal Ltd

Standard Form of Building Contract With Contractor's Design 1998 Edition

Background

During the mid-1970s, a time of popular enthusiasm for industrialised approaches to building, contractor-led design and build became established as an important method of building procurement. It was recommended in the NEDO report, *Construction for Industrial Recovery*. The then Department of the Environment development management working group on value for money in local authority housing also pressed its apparent advantages. The RIBA was asked to raise with the Joint Contracts Tribunal the need for a standard form under which, for a lump sum, a Contractor would design and construct the Works to Employers' stated requirements.

By this time the Department of the Environment and the National Federation of Building Trades Employers (later to become the Construction Confederation) had their own contract forms and fee scales for projects where a building was designed (sometimes using contractor-designed components) to a client's specific requirements. However, no form of contract existed for use by local authorities or the private sector which fairly apportioned the responsibilities, obligations and risks of the parties. Clearly a new form was needed which would deal with the situation where the appointment of neither an architect nor a quantity surveyor was envisaged.

Drafting the new form was protracted, and continued for six years. However, when it was finally published as the JCT Standard Form of Building Contract With Contractor's Design 1981 Edition (JCT WCD81), it was an immediate success and widely accepted within the building industry. Since the form first appeared, public sector housing has declined, and it has been used for building contracts far more complicated than those for which it was originally intended. As the concluding statement in a 1987 report put it: 'Architects should face up to the fact that design and build is here to stay and that many clients appear to welcome the fact'. Design and build is now an established procurement method with many variants, mostly contractor-led but frequently design-led.

The cost of preparing a tender for design and build contracts with a substantial design import is relatively high. The JCT Series 2 Practice Note 6 Main Contract Tendering, gives an excellent framework which should be adopted. If it is to be demonstrably competitive both in design quality and price, then two stage tendering is essential with design and build procurement.

JCT WCD98

Supplementary Provisions introduced in February 1988 greatly increase the usefulness and flexibility of WCD, but will only apply if incorporated by a relevant entry to the Appendix. It should also be noted that Sectional Completion can be incorporated under Article 8, and that the necessary Modifications appear in the printed form after the main body of the Conditions.

Nature

The total document runs to over 120 pages. The Articles of Agreement include six Recitals and eight Articles, and may be executed under hand not as a deed, or as a deed. There are three Appendices: the First dealing with the usual factual information relating to clauses, the Second dealing with alternative basis for payment, and the Third – and very important – a summary of Employer's Requirements, Contractor's Proposals and Contract Sum Analysis on which the agreement rests. Annex 1 to Appendix 1 relates to bonds in respect of advance payments and payment for off-site materials and goods. Annex 2 to the Conditions provides for EDI. There are also supplemental provisions relating to the VAT agreement.

Article 1 states the express obligation of the Contractor as being to 'complete the design for the Works and carry out and complete the construction of the Works' (note that the obligation is to complete the design and not carry out the design). In Article 3 the Employer nominates a person to act as Employer's Agent. This agent may be an architect, surveyor, project manager or any other suitable person, and his or her duties and any limits to his or her authority should be clearly established right from the outset. (The contract wording often refers to the 'Employer' having a duty in certain matters, and depending on the agreement, this duty might or might not rest with the Employer's Agent.) There is no role for a Contract Administrator in the traditional sense.

The Articles of Agreement are prefaced by a headnote. This states that the form is not suitable where the Employer has appointed an architect to prepare or have prepared drawings, Specifications and bills of quantities, etc. In practice, of course, the form is often used where the scheme design has already been developed to a substantial extent, and this headnote is somewhat misleading. Perhaps it should be amended so as to recognise the common practice of appointing an architect to prepare the Employer's Requirements in considerable detail, often including the production of detailed design and even some production drawings.

Although the form bears a marked resemblance to JCT98, and in parts the Conditions are similarly worded, the difference between the forms is fundamental. There is no role for a Contract Administrator to act fairly as between the parties. The basis of the agreement is the compatibility between the Employer's Requirements on the one hand and the Contractor's Proposals on the other.

By the Second Recital, the Contractor is obliged to submit proposals, and a tender figure. The Contractor is also obliged to produce a Contract Sum Analysis. The Third

JCT WCD98

Recital places on the Employer the obligation to examine the Contractor's Proposals. The wording falls short of a warranty by the Employer, but this obligation must be treated cautiously because it is the Contractor's Proposals which prevail in the event of a conflict. (The form is frequently amended to reverse this.) Where the Employer accepts some divergence, the Employer's Requirements should be amended before the contract documents are signed. This does not relieve the Contractor of his obligation to satisfy the Employer's Requirements in terms of design, selection of components and materials, and standards of workmanship – particularly where these are covered by performance specification only.

The Contractor's Proposals should respond to and be consistent with the Employer's Requirements, indicating where amendment or amplification is advisable. They should not include Prime Cost or provisional sums unless the Employer agrees to this, in which case the Requirements must be amended. They should include any necessary plans, elevations, sections and typical details, information about the structural design, services layout drawings, and specifications for materials and workmanship not already provided in the Employer's Requirements, although specifically requested in those Requirements.

The term Employer's Requirements is not really defined, which makes the form equally suitable for projects where there is 'little more than a description of accommodation required', or 'a full "Scheme Design" prepared for the Employer by his own consultants' with planning permission already obtained. (See Practice Note CD1/A.) The Employer's Requirements should state clearly whether the Employer or the Contractor is to be responsible for obtaining approvals. This is a most important matter, and experience suggests that at least outline planning permission is desirable before inviting tenders. If the Contractor is to be responsible for approvals, and these have not been obtained by the date of tender, then under clause 6·3·2, if the Employer's Requirements have stated that any amendments necessary to ensure compliance are not to be regarded as a Change by the Employer, the Contractor may have to bear any extra costs arising. (See JCT Practice Note CD/1A paras 11–13 for further guidance.)

This is a lump sum contract payable in stages or periodically based on the Contractor's valuation. The Conditions make no reference to bills of quantities or a Schedule of Rates. The Contract Sum Analysis will therefore be used for valuing changes in the Employer's Requirements, valuations for interim payments, and for calculating the reimbursement of increased costs by the Formula Rules. The Contract Sum Analysis is to be submitted with the tender, and the Employer's Requirements might stipulate the format and headings to be used, possibly as prepared by the Employer's quantity surveyor consultant.

Use

The form places on the Contractor the same design responsibility as that of an architect or other appropriate professional designer (2·5·1). This of course is to use reasonable skill and care. It is not an absolute liability warranty except to the extent that housing designs must satisfy the provisions of the Defective Premises Act 1972. There is no requirement in the form for the Contractor to take out professional indemnity insurance to back this warranty. Many larger contractors do this as a matter of course, while smaller ones often do not because it is expensive initially, and to be of any value must be maintained at least throughout the liability period.

Liability for consequential loss occurring as the result of design failure by the Contractor may be limited to an amount to be entered in the Appendix (2·5·3). Opinion is divided on the worth of this, and what an appropriate figure might be. As each set of circumstances is different, the Employer should take advice from insurance experts.

The form, as used, sometimes places responsibility for the complete design on the Contractor. It also permits design input by the Employer because the Contractor's design warranty is only applicable 'insofar as the design of the Works is comprised in the Contractor's Proposals'. There can be problems where the Employer has a substantial design input, because in the event of failure the boundaries of responsibility become blurred.

The Employer's Requirements then, may be anything from a simple written statement of performance requirements to a completely developed scheme design, with outline specification and drawings indicating spatial arrangements, materials and finishes, which may have received full planning permission already. However, there is no provision for design input from the Employer after completion of the tender documents, except by way of a variation or 'Change', as this form terms it. Where this occurs the Employer must bear the full cost of such variation, including any consequential expense to the Contractor. Valuation is in the hands of the Contractor.

The fact that there is no place for an architect or Contract Administrator in the contract can leave the problem of quality control unresolved. The Employer's Agent or other person acting with the authority of the Employer or his Agent is to be allowed access to the Works, but there is no inspection in the traditional sense. The Contractor might have engaged his own architect to prepare designs and assist with production drawings, but many Contractors consider the architect's work to be finished once he or she has completed the drawings. Some Contractors choose to retain an architect for site duties to help ensure that drawings are being correctly interpreted, or to see what substitutions can be accommodated without injurious consequential effects. Others regard any questions raised by their architect about workmanship and materials as unwarranted interference, and consider that the inspection of work on site is best left to their site agent and contracts manager. Therefore it might be highly desirable for the Employer to take advantage of clause 11 and appoint an architect or Clerk of Works to inspect and report back, to ensure that the standards of workmanship and the quality of materials is in accordance with the Requirements.

JCT WCD98

Synopsis

1 Intentions

- The Contractor is obliged to carry out and complete the Works referred to in the Employer's Requirements, the Contractor's Proposals, and the printed Conditions. There is express reference to completing the design for the Works, and to reliance upon the Contractor for materials, goods and workmanship otherwise necessary but not referred to in the documents (2·1).
- Kinds and standards will be either those referred to in the documents and if the Employer has specified these in the Requirements, then this will reduce liability on the Contractor. If not, then the Contractor will be wholly liable in the event of failure (2·1 and 8·1).
- The documents are to be read as a whole, and the printed Articles, Conditions, Appendices and Supplementary Provisions prevail in the event of any conflict (2·2).
- The Contractor's Proposals are to be accompanied by a Contract Sum Analysis (Second Recital) and the Employer needs to examine both carefully, because the assumption is that he is satisfied that they meet the Employer's Requirements (Third Recital).
- There is no means within the contract of dealing with a mismatch between the Requirements and the Proposals. Where a discrepancy is found within the Requirements, the Proposals prevail. Where there is a discrepancy within the Proposals, the Employer is to be notified and he must make a decision about the discrepancy and proposed amendments (2·4). This might constitute a Change.
- The Contractor is liable for his own design work to the extent that he warrants reasonable care and skill (2·5). However, where the contract is for housing work which is subject to the terms of the Defective Premises Act 1972, the limit of liability in clause 2·5·1 will not apply (2·5·2).
- The Employer is to be supplied with copies of drawings and other documents prepared by the Contractor, and the Contractor is also obliged to supply 'as built' drawings after completion, and before commencement of the Defects Liability Period (5·3 and 5·5). These may or may not comprise part of the Health and Safety File, according to the circumstances.
- The CDM Regulations require the appointment of a Planning Supervisor and Principal Contractor, and under Article 7 and provision 6A, the CDM obligations on both Employer and Contractor become contractual as well as statutory.

2 Time

- Dates for possession and completion should be entered in the Appendix. The Contractor must proceed regularly and diligently and complete on or before the completion date (23·1).

JCT WCD98

- An option for deferment of possession not exceeding six weeks is available, subject to an Appendix entry (23·1).
- Notice of delay must be given to the Employer, together with supporting information, including the Contractor's estimate of the likely effect on completion (25·2). The Employer is required to notify the Contractor of his decision relating to the completion date within 12 weeks (25·3).
- The listed Relevant Events for which an extension may be awarded includes reference to strikes etc. affecting design work, delay resulting from necessary permissions or approvals, and the effect of changes in statutory requirements or terms of consents which arise after the base date entered in Appendix 1 (25·4). The interim decision by the Employer is subject to review no later than 12 weeks following Practical Completion (25·3).
- If the Contractor fails to complete the Works to time, the Employer may recover liquidated damages, provided that he issues a written notice to the Contractor (24·1).
- Practical Completion, to include sufficient compliance by the Contractor in providing information for the Health and Safety File, is signified by a written statement from the Employer (16·1).
- After this the Contractor is obliged to rectify defects (16·2 and 16·3) unless the Employer decides otherwise and takes an appropriate deduction instead (16·2).
- There is provision for partial possession (17) and for Sectional Completion (Article 8 and Modifications to the Conditions).

3 Control

- The bar to assignment without written consent relates to the contract (18·1). The requirement for written consent to sub-contracting refers not only to the carrying out of work, but also to matters of design (18·2).
- There is an option clause which, if an Appendix entry states that it is to apply, allows the Employer to transfer a right of action against the Contractor to persons with a subsequent interest in the completed Works (including design for which the Contractor was liable) (18·1·2).
- There is no facility for naming or nominating sub-contractors or suppliers in the Conditions, but named sub-contractors may be part of the Employer's Requirements under Supplementary Provision S4.
- Employers' instructions must be in writing, although this can mean written confirmation of oral instructions (4·3). The Conditions clearly define what instructions are empowered and these may include a Change (12) and postponement of any construction work or design (23·2). The Employer cannot order a Change which modifies the design of the Works without the Contractor's consent (12·2).

JCT WCD98

- The Contractor is required to have a competent person-in-charge on the site full-time (10).
- Where work or materials do not comply with the contract, the Employer may instruct the Contractor to remove them from site (8·4). He may also order any consequential Changes necessary, which will not attract any extension of time or addition to the Contract Sum. He may order tests and inspections (8·3) and the likelihood of any non-compliance in similar work elsewhere is covered (8·4·3 and Code of Practice).
- The contract requires all work to be carried out in a proper and workmanlike manner, and in accordance with the Health and Safety Plan (8·1).
- The contract allows for work under the direct control of the Employer to be carried out during the time that the Contractor is in possession (29).

4 Money

- The Contractor is required to provide Priced Statements; there are strict timescales to aid cash flow; and non-payment to time can be a valid reason for suspending work.
- The Contract Sum is VAT-exclusive (14·2) and may only be adjusted as provided for in the Conditions (13).
- The valuation of Employer's Changes can be either by Alternative A – where the Contractor submits a Price Statement (which may include for any consequential effects on completion, direct loss and/or expense), and if this is not applicable or acceptable, by Alternative B – where additional identical or similar work is valued at rates in the Contract Sum Analysis, adjusted if appropriate, and other work which is not measurable on a daywork basis at fair and reasonable rates (12·4 and 12·5).
- The Contractor can recover increases in tax, levies or contributions which arise after the base date (36). Increases in labour and materials may be included for either by the traditional method (37) or by the use of the Formula Rules (38).
- Where provisional sums have been included in the Employer's Requirements, instructions must be given about their expenditure (12·3).
- The Contractor must make written application for reimbursement of loss and/or expense. The grounds for a valid application are set out (26·2) and include loss arising from delay in obtaining planning permission, or other consents which the Employer has undertaken to obtain.
- Interim payments differ from those under other JCT forms. They may be periodic (Alternative B in Appendix 2) or stage payments (Alternative A). The chosen method should be stated in the Employer's Requirements and also, if appropriate, when design costs are to be paid (30·1).
- Applications for interim payment are then made by the Contractor, and each

application must be supported by details as stipulated in the Employer's Requirements.

- The final date for payment is 14 days from the date of receipt of each application. Within five days of receipt, the Employer must notify the Contractor in writing of the amount he intends to pay. Any intention to withhold or deduct money must be stated clearly in a written notice which also sets out the grounds for such action. This must be given to the Contractor no less than five days before the date for final payment. If no valid notices are given by the Employer, then payment in full must be made. Failure to pay amounts properly due by the final date for payment will attract interest on outstanding amounts of 5 per cent over current base rate, and can also give the Contractor the right to suspend work until payment is made (30·3).
- The Contractor has three months following Practical Completion in which to submit his detailed Final Account and the Final Statement (30·5). If the Contractor fails to submit these two important documents within the timescale, the Employer is entitled to prepare his own Final Account and Final Statement and, if not disputed, these will be conclusive as to monetary matters (30·5·6).
- Final date for payment is 28 days from the date when the Final Statement from the Contractor or the Employer becomes conclusive. Failure to make payment in full by the final date for payment will attract interest of 5 per cent over current base rate.

5 Statutory obligations

- The Contractor must comply with all statutory requirements and give all notices required by statute (6·1). The only exception to this contractual obligation is where in the Employer's Requirements it is stated that these are in compliance (e.g. planning permission). All consents or permissions obtained by the Contractor must be passed on to the Employer.
- The Contractor can only claim fees or charges where these are included for in the Employer's Requirements by way of a provisional sum. Otherwise there is no adjustment to the Contract Sum.
- The Contractor is to notify the Employer if he finds any divergence between statutory requirements and either the Employer's Requirements or the Contractor's Proposals. The Employer's consent is required to any necessary amendments (6·1·2).
- If amendments to the Contractor's Proposals become necessary due to changes in statutory requirements after the base date, then these would normally constitute a Change in the Employer's Requirements, and would not be at the Contractor's expense (6·3). However, where the Employer's Requirements expressly preclude this, then necessary amendments would be at the Contractor's expense.
- Where the full CDM Regulations apply, the Contractor is obliged both under statute and in contract to comply with the relevant CDM duties (and this might also be in

respect of design), but in particular concerning the Health and Safety Plan and in providing information for the Health and Safety File (6A).

6 Insurance

- The Contractor indemnifies the Employer in respect of personal injury or death, and damage to property other than the Works (20). This is to be backed by insurance (21).
- If instructed, the Contractor may be required to take out joint names insurance for the Employer against the risk of legal nuisance. This will have been stated in the Employer's Requirements. There is a list of exceptions, and damage must not have been caused by the Contractor's negligence. An Appendix entry on the extent of cover is required (21·2).
- Insurance of 'the Works' follows the JCT98 provisions in clauses 22A or 22B or 22C. The full reinstatement value of work must include for the cost of the Contractor's design work. Which alternative clause is to apply should be stated in the Employer's Requirements and shown by the Appendix deletions.
- In the event that terrorism cover is withdrawn and is no longer available, the situation and options open to the Employer are dealt with in clauses 22A·5, 22B·4 or 22C·5 as applicable.
- Insurance may be required against the Employer's loss of liquidated damages due to an extension of time following damage to the Works (22D). This requires an Appendix entry.
- There is no requirement for the Contractor to take out insurance against the risk of his designers becoming liable for negligence. If the Employer considers (as most do) that this should be provided for, then it must be stated in the Employer's Requirements and included in the Contractor's Proposals. Such insurance cover is the equivalent of professional indemnity cover and is in no way a 'total product guarantee' extending to defective work or materials. Any such cover would need to be maintained for a stipulated minimum period (e.g. up to six years after completion) to be worthwhile.
- An Appendix entry will show if the Joint Fire Code of Practice on the Protection from Fire of Construction Sites is to apply (22FC) and, if so, both Employer and Contractor must comply with it. Any evidence of non-compliance could result in the insurers specifying remedial measures.

7 Termination

- The Employer is allowed to determine the employment of the Contractor by reason of specified defaults (27·2). One reason is where the Contractor suspends carrying out design work and is therefore unable to discharge his obligation to provide drawings. The Employer is first to issue a warning notice, and may follow this with the determination notice. In the case of insolvency of the Contractor, then always

JCT WCD98

depending on the circumstances, determination might be automatic subject to possible re-instatement, or the Employer might enter into an agreement with the Contractor (a '27·5·2·1 agreement') for continuation or novation.

- The Contractor is allowed to determine his own employment for specified defaults. One is delay in obtaining planning permission, where the Contractor is blameless (28). In the event of insolvency of the Employer, the Contractor may elect to determine his own employment.
- Either party can determine the employment of the Contractor for listed neutral causes (28A).
- The respective rights and duties of the parties concerning payment, removal and completion are set out in detail (27·6 and 28·4). Note that the Contractor's obligation extends to providing the Employer with copies of all drawings etc. prepared, and that the Employer's obligation for payment includes design costs.

8 Miscellaneous

- A list of definitions is given (1·3).
- Third party rights under the Contracts (Rights of Third Parties) Act 1999 are excluded (1·9).
- Access for the Employer's Agent, and any person authorised by the Employer, is to be provided (11). This might be subject to reasonable restrictions as far as access to workshops is concerned.
- Discovery of antiquities etc. is covered (34).
- Sums payable in respect of royalties are deemed to have been included in the Contract Sum, and the Employer is indemnified against infringements of copyright (9·1). However, where the Contractor is complying with Employer's instructions (i.e. in matters not covered by the Employer's Requirements) then the Contractor is not liable for infringements, and any additional sums are added to the Contract Sum (9·2).
- The Employer is wholly responsible for defining the boundaries of the site for the Contractor (7). This could be considerably important in the context of the Party Wall etc. Act 1996.
- The Contractor's design warranty includes for consequential loss not covered by liquidated damages (e.g. loss of use, loss of profit, etc.) and can be limited to an amount entered in the Appendix (2·5·3).
- A headnote to the agreement clearly indicates that the form is not suitable for use where drawings, Specifications, bills of quantities have been prepared by the Employer's architect, nor where the Employer wishes to appoint a Contract Administrator.

JCT WCD98

- An appropriate deletion to Appendix 1 should show whether or not the Supplementary Provisions apply. These are seven in total but not every one will necessarily be relevant for all contracts. They can considerably extend the usefulness of the form, and they are as follows:

  S1: number not used. (It provided for mandatory adjudication in 1988, before the Housing Grants, Construction and Regeneration Act 1996 made this a requirement, when the provision became superfluous.)

  S2: where the Contractor is to provide drawings, details and other information, these are to be submitted first to the Employer for comment. The Contractor's liability is not reduced by this requirement.

  S3: the Contractor may be required to appoint a site manager to act as a full-time representative on site, and should a change of person become necessary, the written consent of the Employer is required.

  S4: the Employer is entitled to name a sub-contractor in the Employer's Requirements (note – not to be introduced later during the progress of the Works). The Contractor is to notify the Employer when the sub-contract has been entered into.

  S5: the Employer may include bills of quantities as part of the Employer's Requirements, and if they are firm bills, then the method of measurement used is to be stated.

  S6: where variations or Changes of a major nature have been issued by the Employer, the Contractor may be required to submit estimates of the anticipated effects in terms of cost, extensions of time and consequential expense, before work is authorised. If there is no agreement over terms, then the instruction may be withdrawn, or the matter referred to adjudication.

  S7: where the Contractor makes application for loss and/or expense as provided for under clause 26, the onus is on the Contractor to include a detailed estimate in the application. The Employer may accept the estimate, or choose to negotiate, or refer the matter to adjudication.

9 Disputes

- The Housing Grants, Construction and Regeneration Act 1996 (Part II) gives either party a statutory right to refer any difference or dispute arising out of the contract to adjudication. Article 5 provides for this.
- Procedures for referral to adjudication, the appointment of an adjudicator, the conduct of adjudication etc. are covered in clause 39A.
- The adjudicator's decision is binding on the parties at least until the dispute is finally determined at arbitration or by legal proceedings.

JCT WCD98

- Arbitration is the agreed method for finally settling disputes (Article 6A and 39B). Where in the Appendix this provision is shown as not to apply, then disputes are to be referred for legal proceedings (Article 6B and 39C).

This contract?

If considering using WCD98 remember that:

It is intended for use where the Contractor is to accept responsibility for design of the Works to a greater or lesser extent as the Employer requires, although completion of the design is a stated obligation. The Contractor is to use reasonable skill and care in achieving this (i.e. not a fitness for purpose warranty), and the form does not require the Contractor to hold professional indemnity insurance. The more the Contractor is responsible for design, the clearer the boundaries of design responsibility become. Whilst at first sight, this contract has many similarities with JCT98, a fundamental difference is the absence of any provision for a Contract Administrator or quantity surveyor to act on the Employer's behalf. The form is for use in England and Wales. The Scottish Building Contract Committee publishes a 1999 Edition of the SBC With Contractor's Design form (2002 Revision).

The Employer's Requirements and Contractor's Proposals are the core of this contract and it is important that they are in harmony. The Contract Sum Analysis should be adequately detailed in coverage. Completing the form requires care, particularly because of the number of Appendices and supplemental and supplementary provisions available.

The Employer would be well advised to consider incorporating the optional Supplementary Provisions, particularly those relating to Contractor's submission of drawings, named persons in Employer's Requirements, and submission of estimates by the Contractor relating to the valuation of Changes and loss and expense.

If acting as Employer's Agent, any limits to authority should be clarified, and a clear understanding reached on what is empowered by the Employer under the contract. If acting as consultant advising the Employer, care is needed to stay strictly within the limits of the appointment especially once work starts on site. If acting for the Contractor under a novation agreement, accountability should be clearly established and respected.

This was the first agreed standard form exclusively for design and build. Obviously adapted from JCT98 and with the same lack of logical structure. The wording is for the most part familiar, and care is needed not to overlook important differences in some provisions. However, the form appears to have worked well enough in practice, and given the current enthusiasm for design and build procurement its popularity will undoubtedly continue.

JCT WCD98

Related matters

Documents

Standard Form of Building Contract With Contractor's Design
1998 Edition (WCD98)
Amendment 1: 1999 (Construction Industry Scheme)
Amendment 2: 2000 (sundry amendments)
Amendment 3: 2001 (terrorism cover/Joint Fire Code/CIS)
Amendment 4: 2002 (extension of time/loss and expense/advance payment)
Sectional Completion Supplement is incorporated in WCD98

References

Earlier JCT Practice Notes (relevant to WCD98 but written with JCT80 in mind):
Practice Note No 23: Contract Sum Analysis (1987)

Series 2 JCT Practice Notes (yellow covers):
Practice Note 1: Construction Industry Scheme
Practice Note 2: Adjudication (includes text of agreements)
Practice Note 3: Insurance, Terrorism Cover
Practice Note 4: Partnering (includes text of non-binding charter)
Practice Note 5: Deciding on the Appropriate JCT Form of Contract
Practice Note 6: Main Contract Tendering (includes model forms)

JCT Practice Note CD/1A (written for WCD 1981 but still largely relevant)
JCT Practice Note CD/1B (written for WCD 1981 but still largely relevant)

Commentaries

David Chappell and Vincent Powell-Smith
The JCT Design and Build Contract
2nd edn, Blackwell Science (1999)

Sarah Lupton
Guide to WCD98
RIBA Enterprises (2002)

Dennis Turner
Design and Build Contract Practice
2nd edn, Longman (1996)

JCT CDPS

The Joint Contracts Tribunal Ltd

Contractor's Designed Portion Supplement 1998

for use with the Standard Form of Building Contract 1998 Edition

Background

There is often a design in traditional procurement contracts, which in practice is left to the Contractor or sub-contractors. Strictly speaking, unless this is covered by appropriate wording in the contract, it might be argued that such responsibility lies outside the contract obligation to carry out and complete the Works.

Some contract forms provide for this. Others, like many JCT forms, accept that some design by named or nominated sub-contractors may be inevitable, but clearly state that the Main Contractor is not to be held responsible. This usually means that a collateral agreement between the sub-contractor and the Employer is desirable.

With some projects for which JCT98 is used, it might be desirable to go beyond the usual 'work and materials' concept and make the Main Contractor responsible for specific design matters. This might be in respect of some discrete portion of the building, or more likely for some part where there could be a proprietary system element. This should be only for an identified part of the Works, and under no circumstances should this approach be adopted where the proper answer lies in the use of a full design and build form of contract.

To provide an answer in such situations, the JCT publishes a Contractor's Designed Portion Supplement (CDPS) for use with JCT98 by which design responsibility for certain defined areas of work can be placed with the Main Contractor. This Supplement is only for use with JCT98, and is available in a With Quantities or Without Quantities version.

In the Contractor's Designed Portion Supplement, the Contractor does not warrant fitness for purpose and is only required to exercise reasonable care and skill.

Nature

The Contractor's Designed Portion Supplement contains substitute Articles of Agreement, Recitals, an extended Article 1 and a Supplementary Appendix. As it constitutes a Supplement to the main document, there is no need for separate attesting.

As with the JCT WCD98 form, Employer's Requirements, Contractor's Proposals, and a CDP Analysis for the defined portion are essential requirements to make this Supplement work.

JCT CDPS

A Schedule of Modifications imports the necessary substitutions, deletions and additions to the Conditions in the main JCT98 form.

Use

First check that the appropriate version of Contractor's Designed Portion Supplement is being used, compatible with the JCT98 form of contract relating to revisions and amendments. There are versions of the CDPS for use With Quantities and Without Quantities. There is also a choice between a document solely for sectional completion, or the Composite Contractor's Designed Portion and Sectional Completion Supplement, in either With Quantities or Without Quantities versions. This latter option results in a considerable number of Modifications and some arduous cross-referencing, but at least it should ensure proper integration.

The Articles of Agreement, Recitals and Supplementary Appendix must be completed, and new Article 1 and Modifications to the Conditions will be incorporated.

If using the Contractor's Designed Portion Supplement it should be remembered:

- Whilst the Architect can notify the Contractor of anything in the design which 'appears' to be defective, even if no such notice is given, it does not relieve the Contractor of his obligations to prepare and complete the design (2·9), nor does it affect the Employer's option regarding the limitation of the Contractor's liability for consequential damage.
- Whilst the Architect has power to issue directions to the Contractor to make certain that his designed portion can be integrated into the design of the Works as a whole (2·1·2), the Contractor has the right to object if he considers that this would affect the efficacy of his own design (2·8).
- In response to the Employer's Requirements the Contractor is obliged to supply Contractor's Proposals and a CDP Analysis for that part of the Contract Sum which relates to the Contractor's Designed Portion. The latter will assist the valuation of variations (Fourth Recital).

Synopsis of Modifications

1 Intentions

- The Contractor's Designed Portion Supplement (CDPS) is to be incorporated into the Standard Form of Building Contract (JCT98), either the With Quantities or Without Quantities version, which then becomes a modified agreement.
- The additional obligation on the Contractor to complete the design is restricted to the part of the Works identified. The integration of this design with the design for the Works as a whole is the responsibility of the Architect (2·1·2).
- The Employer's Requirements may include drawings and bills of quantities and other

JCT CDPS

documents. The Contractor's Proposals for the CDPS must be accompanied by a CDP Analysis relating to these. The Requirements, Proposals and Analysis only relate to the identified portion.

- The provision found in JCT98 for dealing with a discrepancy or divergence is extended to include the Employer's Requirements, the Contractor's Proposals and the Analysis (2·3).
- The Contractor's liability for design is limited to exercising reasonable care and skill (2·7·1). However, where the contract is for housing work which is subject to the terms of the Defective Premises Act 1972, then this limit of liability might not apply.
- The Contractor must be notified if the Architect detects a defect in the Contractor's design, and the Architect must be notified if the Contractor considers that his design will be adversely affected by an Architect's instruction. The Contractor's obligations for design work are otherwise not reduced (2·9).

2 Time

- No extension of time is empowered because of delays arising out of design information to be provided by the Contractor for the portion (2·10).

3 Control

- Materials, goods and workmanship for the portion are to be as described in the Employer's Requirements. The Contractor is not permitted to make any substitutions without obtaining the Architect's written consent (8·1).
- If the Architect issues a variation on work which is part of the portion, then this is a modification of the Employer's Requirements and will be valued as such (13·2 and 13·8).
- The Contractor is only permitted to sub-contract design work for the portion with the Architect's written consent and is nevertheless still liable for that design.

4 Money

- The Analysis provided by the Contractor is used as a basis for valuing variations to the portion (13·8).

5 Statutory obligations

- If divergences between statutory requirements and documents for work on the portion are discovered, the Contractor is to propose the necessary amendments, to be at his own cost (6·1·8).
- CDM Regulations requirements in respect of design by the Contractor in the portion, as well as other relevant statutory matters, are contractual as well as statutory obligations.

JCT CDPS

6 Insurance

- The CDPS does not require the Contractor to take out any insurance in respect of design liability.

7 Termination

- The CDPS introduces only minor modifications, but failure to proceed regularly with design work in respect of the portion is included as a reason for determination by the Employer (27·2). In the event of determination, the Contractor is to provide the Employer with completed design information (27·6).

8 Miscellaneous

- The Contractor's liability for consequential loss arising due to design errors not covered by liquidated damages may be limited to a figure entered in the Supplementary Appendix (2·7·3).
- The Final Certificate of the Architect is not intended to be conclusive in respect of work, materials, goods or any design completed by the Contractor for the Contractor's Designed Portion (30·10).

9 Disputes

The relevant standard form clauses are not modified, and will therefore provide for adjudication and either arbitration or legal proceedings.

JCT CDPS

This contract?

If considering using CDPS with JCT98 remember that:

It is intended for use only with JCT98 in the With Quantities and Without Quantities versions. It is available as a separate Supplement, or as a Composite Supplement which also includes for Sectional Completion. It does not make JCT98 suitable for design and build, but it makes the Contractor responsible for the design of an identified portion of the Works. It was not intended as a means of placing design responsibility on the Contractor for work sub-contracted under JCT98 clause 19·3.

The Supplement introduces substitute Articles, Modifications and a Supplementary Appendix. Limits of responsibility need to be defined. It is important to remember that the Architect is still responsible for the integration of any Contractor's Designed Portion work into the overall design and must issue empowered instructions as necessary in order to achieve this.

The fundamental difference between a contract modified by CDPS and a With Contractor's Design contract is that in the former control still rests largely with the Employer through his appointed Contract Administrator and quantity surveyor. These persons are not present under the terms of most design and build contracts. Design may be introduced into JCT98 by means of nominated sub-contract work for which the Contractor accepts no design responsibility; performance specified work which usually has some element of design; or by means of the CDPS. Which option is likely to prove more appropriate for a given situation will depend on the circumstances, and a decision requires careful thought.

Related matters

Documents

Contractor's Designed Portion Supplements (for use with both Private and Local Authority versions of JCT98):
Contractor's Designed Portion Supplement With Quantities 1998 Edition (revised 2001)
Contractor's Designed Portion Supplement Without Quantities 1998 Edition (revised 2000)
Composite Contractor's Designed Portion and Sectional Completion Supplement With Quantities 1998 Edition (revised 2000)
Composite Contractor's Designed Portion and Sectional Completion Supplement Without Quantities 1998 Edition (revised 2000)

References

JCT Practice Note CD/2: Contractor's Designed Portion Supplement (Revised 1996)

ICE/D&C

The Institution of Civil Engineers

ICE Design and Construct Conditions of Contract Second Edition (2001)

Background

In the early 1990s government moved to adopt design and build as the basis for awarding contracts for road schemes costing up to £40 million. It was a further endorsement of the rapidly growing popularity of this method of procurement, and the belief that there would be a 'greater incentive to complete schemes more quickly and save money overall – even though the up-front costs may be higher'.

The ICE Design and Construct contract was therefore a timely introduction in 1992, although obviously closely modelled on the traditional ICE Conditions The structure of the two forms is broadly similar, even down to the same section headings. There, however, the similarity ends because in nature and purpose the documents are quite distinct.

As with the traditional ICE Conditions, this form is produced by the Institution of Civil Engineers, the Association of Consulting Engineers and the Civil Engineering Contractors Association, and is kept under regular review.

Nature

The document is 70 pages long, but this includes the Conditions, a Form of Tender with a two-part Appendix, a Form of Agreement, a Form of Default Bond, and Fluctuation clauses.

Contractor's design obligations are for both permanent and temporary works. The Form of Agreement will be completed to show which documents are intended to be part of the contract. Among these the Employer's Requirements and the Contractor's Submission are key essentials.

The Conditions of the contract are in 72 clauses, and the sequence follows closely that which is found in the traditional ICE Conditions. The Contents list and Index on a clause by clause basis are particularly helpful.

Use

Entries in Part 1 of the Form of Tender Appendix will show a Contractor tendering whether quality assurance or a performance bond is required, and provide detailed information such as commencement date, time for completion, completion by sections, damages, payment provisions, CDM responsibilities and arbitration procedures. Part 2 of the Appendix will be completed by the tenderer.

ICE/D&C

The key person in terms of contract administration is the 'Employer's Representative'. He or she is to be named in the Appendix Part 1, and is given considerable authority to act within the terms of the contract. In the event of his or her departure the Employer must nominate a replacement.

Synopsis

1 Intentions

- The Contractor undertakes to design, construct and complete the Works, including providing all design services, labour and materials – everything whether of a permanent or a temporary nature as specified in or reasonably to be inferred from the contract (8).
- In all design obligations the Contractor exercises all reasonable skill, care and diligence (8) and this includes accepting responsibility for design work included as part of the Employer's Requirements.
- The Contractor is to institute a quality assurance scheme to an appropriate extent, which must be approved by the Employer's Representative before commencement of work at both design and construction stages (8).
- The Contractor is responsible for the safety of the design, and for stability and safety of all site operations and methods of construction (8).
- Definition of the Works includes both temporary and permanent work. Contract means Conditions of Contract, Employer's Requirements, Contractor's Submission and other documents as agreed by the parties (1). The documents taken together are stated to be mutually explanatory, but in the event of discrepancies between the Contractor's Submission and the Employer's Requirements it is the latter which prevail (5). (This is the reverse of JCT WCD98.)
- The Contractor is deemed to have inspected and examined the site and surroundings, and the information provided by the Employer before tendering (11). In the event of physical conditions or artificial obstructions which could not reasonably have been foreseen, then provided that the Contractor gives written notice to the Employer's Representative, additional payment may be authorised (12).
- The Contractor must submit all necessary design drawings to the Employer's Representative and his or her consent must be obtained before construction work is undertaken (6).
- The Contractor must submit a programme to the Employer's Representative for his or her acceptance. It must be revised from time to time as necessary (14).
- The Works must be designed, constructed and completed in accordance with the contract, and materials and workmanship must be as described in the contract, or be to appropriate standards and codes of practice. There is provision for checks and testing to be carried out.

ICE/D&C

- Reasonable facilities must be provided for any other contractors employed by the Employer on or near the site of the Works (31).
- The functions and authority of the named Employer's Representative and his or her named assistants are clearly defined (2).

2 Time

- The Commencement Date will be as stated in the Appendix to the Tender, or as agreed by the parties, or otherwise 28 days after entering into the contract (41).
- The Contractor must start as soon as reasonably practicable, and proceed with due expedition and without delay (41).
- The contract may prescribe that possession of the site will be in portions, and also determine the order of availability and order of the Works (42).
- Failure by the Employer to give possession which results in delay to the Contractor can result in the award of extensions of time and expense (42).
- Extensions of time may also be due for reasons of ordered variations, weather or special circumstances of any kind (44). The Contractor should notify the Employer's Representative within 28 days of a delay and supply necessary particulars. The Employer's Representative is empowered, not later than 14 days after the completion date, to award an extension of time even if there is no claim by the Contractor.
- If the Contractor fails to complete the whole or any designated section of the Works to time, the Employer may deduct and retain liquidated damages. These must not be a penalty, but where no limitation to liquidated damages is stated in the Appendix to the Form of Tender, then liquidated damages without limit shall apply (47).
- Where the Employer's Representative requests accelerated completion and the Contractor agrees, then the terms for payment shall be agreed between the parties before action is taken (46).
- Where progress is too slow to ensure completion by the date agreed, the Employer's Representative may notify the Contractor of his or her opinion, and the Contractor is obliged to take such steps as may be necessary at his own expense (46).
- The Employer's Representative must issue a Certificate of Substantial Completion when in his or her opinion the whole or a designated section of the Works is substantially completed. The Contractor must notify the Employer's Representative when completion has been achieved, and the Employer's Representative issues the certificate within 21 days of notification, always provided that he or she is satisfied (48).
- The Contractor is obliged to complete any outstanding work and deal with repairs during the Defects Correction Period. When the Employer's Representative is

ICE/D&C

satisfied, a Defects Correction Certificate is issued, although this in no way relieves the Contractor of any liability (61).

- Prior to the issue of the Defects Correction Certificate, the Contractor must submit manuals and as-built drawings for the permanent works. This is a contractual obligation quite additional to material for the Health and Safety File (61).

3 Control

- Neither party may assign the contract or any benefit or interest under it without the written consent of the other (3).
- The Contractor must first obtain consent from the Employer before making any change of the Contractor's designer from the person named in the Appendix to the Form of Tender, Part 2 (4).
- The Contractor is permitted to sub-contract any part of the construction work, but must notify the Employer's Representative prior to the named sub-contractor arriving on site (4).
- The Employer's Representative may order suspension of any part of the Works (40) and order alterations to the Employer's Requirements (51).
- The Contractor is obliged to give the Employer's Representative full opportunity to inspect work before it is covered up (38).
- The Employer's Representative may instruct the removal from the site of any materials which do not comply with the contract, and the removal and replacement of materials and workmanship. This will extend to replacement of work for which the Contractor has design responsibility (39).
- The Contractor must provide all necessary superintendence. A Contractor's Representative will have overall responsibility and may delegate to a nominated deputy subject to the agreement of the Employer's Representative (15).

4 Money

- The Form of Tender includes for a lump sum, or such other sum as may be ascertained in accordance with the contract Conditions. The contract price will include for the design, construction and completion of the Works (1).
- VAT will not have been included in the contract price (70).
- Provisional sums and Prime Cost sums are not referred to in the contract, but a 'Prime Cost Item' means a Prime Cost sum for the supply of goods, materials or services (1). Use of Contingency and Prime Cost Items require prior consent by the Employer's Representative.

ICE/D&C

- Variations to the Employer's Requirements may be ordered, and if requested by the Employer's Representative the Contractor must submit an estimate of the extra cost and delay involved. Otherwise, or if the estimate is not accepted, valuation of ordered variations will be by the Employer's Representative on a fair and reasonable basis and in accordance with the contract (52). Work ordered on a daywork basis will be valued in accordance with the Federation of Civil Engineering Contractors Schedule of Dayworks (55).
- Payment to the Contractor is made on the basis of interim statements submitted to the Employer's Representative as prescribed in the contract, followed by an Interim Certificate within 25 days. The Certificate must show the amount due and the basis of calculation. If the Employer intends to withhold payment, he must notify the Contractor not less than one day before the final date for payment, and state the grounds for any deduction. Final date for payment by the Employer is 28 days after delivery of the Contractor's monthly statement to the Employer's Representative. In the event of failure to certify, or failure on the part of the Employer to make proper payment to time, the Contractor is entitled to interest on the outstanding amounts (60).
- Retention amounts will be shown in the Appendix to the Form of Tender, and the payment of retention will be subject to the issue of the Certificate of Substantial Completion, and the end of the Defects Correction Period (60).
- Not later than three months after the date of the Defects Correction Certificate, the Contractor must give the Employer's Representative a final account together with supporting documents. The Employer's Representative then has three months to verify this and to issue a Final Certificate. Final date for payment by the Employer is within 28 days of certification (60).

5 Statutory obligations

- The Contractor must give all notices and pay all fees required by legislation. This might be in respect of design or construction relating to both temporary and permanent work. If the Employer's Representative certifies this, then the Contractor can expect reimbursement (26).
- The Contractor indemnifies the Employer against the consequences of any breach of statutory obligations, but this will not apply if this arises due to complying with an instruction given by the Employer's Representative. The Contractor is not responsible for obtaining planning permission unless the contract actually requires this. If the Employer's Requirements do not conform with statutory requirements, then the Employer's Representative must issue necessary corrective instructions (26).
- The CDM Regulations 1994 place obligations on both Employer and Contractor, particularly in respect of the Health and Safety Plan and the Health and Safety File. The Employer is obliged to appoint a Planning Supervisor and a Principal Contractor. With the incorporation of these provisions into the contract, these statutory duties become contractual obligations as well (71).

ICE/D&C

6 Insurance

- The Contractor takes full responsibility for the care of the Works, materials, plant and equipment from Commencement Date to Substantial Completion. The risks include any loss or damage from whatsoever cause, but do not include the Excepted Risks listed in the contract Conditions (20).
- The Contractor must insure in joint names against risk of damage to both the temporary and permanent works for the full reinstatement cost, plus 10 per cent to cover additional costs (21). The terms of all insurances are for approval by the Employer.
- There appears to be no requirement for the Contractor to take out insurance cover in respect of design failure.
- The Contractor indemnifies the Employer against the consequences of claims for injury to persons and damage to property other than the Works. There are stated exceptions, and these remain the responsibility of the Employer (22).
- The Contractor is required to cover the indemnity afforded the Employer by taking out third party insurance cover. The minimum cover required by contract will be the figure stated in the Appendix to the Form of Tender, but this will not necessarily be the extent of the Contractor's liability (23).
- The Contractor's obligations concerning accident or injury to work people is reinforced, and the Employer is indemnified in respect of claims (24).

7 Termination

- The Employer may give the Contractor seven days' notice in the event of specified defaults, which include insolvency, and may expel the Contractor without releasing him from any obligations under the contract (65).
- Procedures for completing the Works, ascertaining the value of work already done, and arranging for payments after termination, are set out in the contract (65).
- In the event of specified defaults on the part of the Employer, which include insolvency, the Contractor may terminate his own employment under the contract after serving seven days' notice. If default continues for a further seven days the Contractor may with all reasonable dispatch remove all equipment from the site (64).
- Following termination by the Employer, the Conditions provide for assignment of goods and materials, but do not appear to include for design drawings which might be necessary to complete the Works.

ICE/D&C

8 Miscellaneous

- The Conditions include a full set of definitions (1).
- The Employer's Representative can require the removal of employees of the Contractor (16).
- There is provision in the event of discovery of fossils, antiquities and things of archaeological interest, etc. (32).
- The Employer's Representative is to have access to the Works, the site and the workshops. The Conditions refer to work in preparation, materials in manufacture, but not to offices where design information is being prepared (37).
- Matters relating to the possession of site, and site access, beyond those prescribed in the contract are mainly the responsibility of the Contractor. Where there is a failure to give the Contractor possession on time, extension of time and additional costs can be awarded (42).
- Contractor's plant and equipment, goods and materials brought on to the site may not be removed without the written consent of the Employer's Representative (54).
- In the event of an outbreak of war, the rights and obligations of the parties are as stated (63).
- Rights of third parties under the Contracts (Rights of Third Parties) Act 1999 are excluded (3).
- Special conditions may be incorporated, and they should be numbered consecutively after the standard Conditions of contract (72).

9 Disputes

- The terms of the contract require that matters of dissatisfaction must be referred to the Employer's Representative in the first instance. The Employer's Representative is required to state his or her decision in writing to both the Employer and the Contractor within one month of the reference.
- If a matter of dissatisfaction cannot be satisfactorily resolved by reference to the Employer's Representative, then it becomes a dispute. The parties might, by agreement, seek consideration of the dispute under the ICE Conciliation Procedure (1994).
- Since the Housing Grants, Construction and Regeneration Act 1996 (Part II) came into force, all construction contracts carry a provision, either express or implied, that parties to the contract have the right of reference to adjudication. With the ICE Conditions of Contract a notice of adjudication is to be given, and adjudication is to be conducted in accordance with the ICE Adjudication Procedure (1997).

ICE/D&C

- All disputes may be finally determined by reference to arbitration. The party seeking arbitration must serve a notice to refer. Arbitration is to be under the Arbitration Act 1996, and conducted in accordance with either the ICE Arbitration Procedure (1997) or the Construction Industry Model Arbitration Rules.

This contract?

If considering using the ICE/D&C remember that:

This form is intended for design and build by the Contractor. It is primarily for use with civil engineering work. It may be for a lump sum or measurement. The Employer is required to appoint an Employer's Representative to act as contract administrator.

The contract is for use under the law of England and Wales, and is also suitable for use in Northern Ireland or under Scots law as provided for in clause 67.

The key features are the Employer's Requirements, and the Contractor's Submission. Unlike the JCT WCD98, in the event of any conflict, it is the Employer's Requirements which take precedence. The onus is on the Contractor to check design information supplied as part of the Employer's Requirements. Drawings originating from the Contractor's designers must be approved by the Employer's Representative before work is commenced, but this in no way reduces the Contractor's liability.

This is a welcome and significant design and build contract. For civil engineers, the change from traditional role to Employer's Representative is likely to be much easier than that facing architects acting under the Conditions of the JCT WCD98 contract.

Related matters

Documents

ICE Design and Construct Conditions of Contract Second Edition (2001)

GC/Works/1 Design Build

The Stationery Office

GC/Works/1 Single Stage Design and Build (1998)

GC/Works/1 Two Stage Design and Build (1999)

Background

GC/Works/1 Edition 3 for traditional procurement was first published in 1989. Shortly afterwards, with the demise of the Property Services Agency, government departments assumed responsibility for their own projects. Some looked to non-traditional procurement approaches, design and build in particular, which was becoming increasingly popular. In response to this interest, a version of GC/Works/1 for Single Stage Design and Build was introduced in 1993 and is currently in a 1998 revised form. A 1999 version for Two Stage Design and Build has subsequently been published.

The Single Stage Design and Build version is obviously an adaptation of the traditional procurement GC/Works/1 contract. It is mainly a lump sum contract and does not distinguish a separate design phase. It is flexible in that it allows for varying amounts of design input by the Contractor, as the Contractor responds to the Employer's Requirements by developing the design outlined in these documents.

The Two Stage Design and Build version is also mainly a lump sum contract, but where the design may not be sufficiently advanced to enable the Contractor to submit a realistic tender, a design fee is submitted initially together with a Schedule of Rates. This is used to quantify the construction price at the end of the design phase. Completion of the design phase will not necessarily lead to the Employer proceeding with construction.

Nature

For the purposes of this book, comments will in the main focus on the Single Stage version of the design and build contract.

In common with other GC/Works forms, this comes as a two volume pack: first the General Conditions which run to over 80 pages, and secondly the Model Forms and Commentary. The General Conditions are prefaced by a Contents list, followed by an Index. Language and terminology is as that to be found in other GC/Works/1 contracts. The 65 clauses appear under the standard nine headings, and a large

GC/Works/1 Design Build

number are identical with those found in the traditional form. There is also a very useful Schedule of Time Limits, the essential Abstract of Particulars with an Addendum which is really a schedule of design information, an Invitation to Tender, Tender and Tender Price Form, and the Contract Agreement.

Use

The Single Stage version is without a separate design stage. The design input required from the Contractor will be to the extent desired, and as indicated in the Employer's Requirements. The Contractor's Proposals will be submitted as part of the Tender together with a Programme, a Pricing Document, and details of professional indemnity insurance.

The Tender Price Form includes alternative entries depending on whether, in Condition 10, Alternative A (design liability limited to using reasonable care and skill), or Alternative B (warrant of fitness for purpose) applies.

The Two Stage version calls for a separate design stage, and the lump sum figure tendered is arrived at in two stages.

The Conditions in GC/Works/1 Design and Build provide for:

- design documents, with copyright in design and documents established;
- professional indemnity insurance for design;
- incentive bonus for early completion;
- finance charges;
- mobilisation payments;
- payments to the Contractor on the basis of stages, milestones, or valuations;
- performance bonds;
- parent company guarantee;
- collateral warranties;
- as built drawings and documents.

The factual details relating to a particular contract, and the incorporation of option provisions will be determined by how the Abstract of Particulars is completed. The Abstract is detailed and amongst other things requires the names of the Project Manager and Planning Supervisor (who may be the Project Manager). The provision of a Project Manager acting on the Employer's behalf, and given so much authority, is somewhat unusual in design and build contracts. The adjudicator and the arbitrator may also be named.

Synopsis (Single Stage version)

The form is one of the family of GC/Works/1 contracts. The Conditions are broadly similar to those of the traditional GC/Works/1 Form, and it is therefore unnecessary to repeat much of which appears earlier in Chapter 6 above.

GC/Works/1 Design Build

However, the design and build forms differ in several important respects, some of which are as follows:

- The definitions include items mainly relevant to design and build, such as Employer's Requirements, Contractor's Proposals, Pricing Document, design, Design Document, etc.
- The fair dealing and team-working obligation extends to the project team, including those responsible for design and costs (1A).
- In the event of discrepancy between Employer's Requirements and Contractor's Proposals, it is the Requirements which prevail (2[2]). This is the reverse of the position with JCT design and build documents.
- The professional indemnity insurance requirements relating to design are similar for both the traditional and design and build versions (8[A]), although in the former case these would apply only if stated in the Abstract of Particulars.
- The Contractor is solely responsible for the correctness of setting out, and there is also a requirement on the Contractor to supply full 'as built' drawings and other relevant information within 14 days of the Date of Completion (9).
- Although a Contractor's design obligation can be incorporated into the traditional form, the design obligations are slightly different from those for design and build (10).
- The Design Documents provision (10A) is peculiar to the design and build form.
- The provision on Foundations (16) in the traditional form is not included in the design and build version.
- The Contractor must provide samples as are specified in the Employer's Requirements and obtain approval before commencing work (31[3]).
- The acceleration provisions differ (38).
- In the sub-letting provisions, the design and build form makes no reference to nominated sub-contractors (62). Obviously therefore, the nominated sub-contractor provision (63) in the traditional form is not used.

GC/Works/1 Design Build

This contract?

If considering using GC/Works/1 Design and Build remember that:

Although there is a Contractor's design provision in the traditional GC/Works/1, this is the true design and build version. Depending largely on the design information contained in the Employer's Requirements, a choice between single stage or two stage tendering will determine which version of the design and build form is most applicable.

The now defunct National Joint Consultative Committee, and the present Construction Industry Board, have both produced excellent Codes for the selection of design and build contractors. Both advocate that design and build tendering is best achieved through the two stage process, and state that single stage tendering will be suitable only where the Employer's Requirements are for a well defined design with little or no risk of further modification.

When completing the contract details in the GC/Works forms, the Abstract of Particulars is a key document. If any special supplementary conditions are incorporated, then in the event of conflict these prevail over the printed Conditions. This, of course, is the reverse of the position with JCT contracts.

The wording of the Conditions is clear and well presented. There are alternatives for design liability depending on whether this is for the professional duty to use reasonable care and skill, or for an absolute fitness for purpose.

Contract administration should be relatively straightforward and there is the customary GC/Works provision for progress meetings in Condition 35. The Model Forms are published in a separate supporting document and must be used. There are eight documents collateral to the contract, and a further 13 administration forms.

The government has recently set departments targets to become best practice clients. These cover the notions of integrated supply chain routes such as design and build, and value for money, taking into account whole life costing and value management. These 'Achieving Excellence' targets may be progressed by incorporating Amendment 1 into the Abstract of Particulars for GC/Works/1 Design and Build contracts.

This form will appeal most to those with experience of GC/Works/1, for the terminology and procedures will be familiar. It is particularly interesting because it seems to give the Employer and the Project Manager a degree of control over the Contractor not usually found with design and build contracts.

GC/Works/1 Design Build

Related matters

Documents

GC/Works/1 Single Stage Design and Build (1998) General Conditions
GC/Works/1 Single Stage Design and Build (1998) Model Forms and Commentary
GC/Works/1 Two Stage Design and Build (1999) General Conditions

Amendment 1: Achieving Excellence

Management procurement 12

Management forms

The Joint Contracts Tribunal Ltd
Standard Form of Management Contract 1998 Edition

The Joint Contracts Tribunal Ltd
Works Contract 1998 Edition for use with the Standard Form of Management Contract

The Joint Contracts Tribunal Ltd
Construction Management Agreement C/CM

The Joint Contracts Tribunal Ltd
Trade Contract TC/C for use with the JCT Client and Construction Manager Agreement

It is accepted that several major forms of building contract for traditional procurement can be adapted for use in management contracting. However, the forms included in this chapter are those exclusively for use in contracts where the principal role of the Contractor is to manage the intended works, which are carried out by other persons under his or her control.

JCT MC98

The Joint Contracts Tribunal Ltd

Standard Form of Management Contract 1998 Edition

Background

For 'fast-track' projects where the Employer still wants the overall design, Specification and contract administration left in the hands of an independent professional team, management contracts are one solution. Their use in the United Kingdom became popular during the 1980s, but in recent years they seem to have lost ground to construction management. Major client bodies have become more sophisticated and well able to handle the direct involvement associated with this latter type of procurement.

In 1979 the RIBA Council, on the advice of its Contracts Committee, asked the JCT to produce a standard form of management contract. At the time, the only forms available were those devised by contracting organisations who pioneered this kind of working. These were often geared to suit the preferred working procedures of the companies, and understandably drafted with their particular interests very much in mind.

In 1987 the JCT issued the Standard Form of Management Contract (MC87) together with related documents necessary for management contracting. The main documents were the head contract between the Employer and the Management Contractor, and Works Contracts between the Management Contractor and each 'Works Contractor' carrying out a package of the work.

Nature

The documents are currently published in 1998 Editions and the Management Contract is in familiar JCT format with section headed Conditions, whilst the Works Contract is published in three discrete parts, the first of which is itself in three parts namely:

- Works Contract/1, Section 1: Invitation to Tender
- Works Contract/1, Section 2:Tender
- Works Contract/1, Section 3: Articles of Agreement (to which Sections 1 and 2 are annexed)
- Works Contract/2 (sub-contract conditions, which may be incorporated by reference)
- Works Contract/3 (Employer/Works Contractor agreement)

The main contract between Employer and Management Contractor covers both the pre-construction period and the period of actual construction work. It contains a detailed Contents list and the Articles of Agreement which follow include Recitals, Articles and provision for attestation. The Conditions are followed by an Appendix in

JCT MC98

two parts, Part 1 relating to information required generally and Part 2 relating to specific information which needs to be entered before construction work starts. Necessary to this kind of document are the Schedules which appear immediately after the VAT agreement.

The First Recital refers to the description of the project as entered in the First Schedule, and confirms that the Employer has appointed a professional team. The Third Recital confirms that the Management Contractor is to cooperate with the professional team, both in the Pre-Construction Period and the Construction Period in respect of services set out in the Third Schedule.

Article 1 confirms the Management Contractor's agreement to perform the services defined in the Third Schedule, for the amounts which the Employer agrees to pay under Article 2.

The names of the Architect, the Contract Administrator and the quantity surveyor are to be entered in Articles 3A, 3B and 4. Other members of the professional team are to be entered in Article 5.

In Article 6 the Employer undertakes to have Project Drawings and the Project Specification, and a Contract Cost Plan prepared as soon as reasonably practicable after the date of the contract. By Article 7, the Employer undertakes to have necessary information prepared by the professional team in respect of the Works Contracts.

The appointments of the Planning Supervisor and Principal Contractor for the purposes of the CDM Regulations are covered in Article 9. Article 8 confirms the rights of either party to refer disputes to adjudication, while Article 10 covers whether the final resolution of disputes is to be by arbitration or by litigation.

The contract may be executed under hand and not as a deed, or as is more likely with this type of contract, as a deed.

Use

The Management Contractor may be appointed by the Employer pre-construction, at a stage early enough to be able to contribute to the work of the professional team. For this he will expect to receive a fee. Then, assuming that the project proceeds to the construction period, the Management Contractor will appoint Works Contractors to carry out the 'work packages'. For this he will be paid a management fee and be reimbursed the Prime Costs as defined in the Second Schedule.

Management Contractors are almost invariably selected by tender and after interview. The fee is not usually the main criterion; this is above all a contract about resources and the ability to manage effectively many Works Contractors.

The Management Contractor will advise on the choice of the Works Contractors. They will normally be appointed on the basis of competitive tendering. The Management

JCT MC98

Contractor starts with a Contract Cost Plan and Programme dates. He will be responsible for the appointment of Works Contractors, their coordination, supervision, and the provision of all site services and facilities.

Tight financial control is essential, and considerable reliance is placed upon the ability of the Management Contractor to monitor the Cost Plan total, even though an independent surveyor is appointed by the Employer. The Management Contractor is under a contractual obligation to achieve completion on time, although any design developments or detailed changes in work packages which occur as the work proceeds could give rise to extensions of time. The contract still requires the Management Contractor to use 'best endeavours' and also allows for acceleration of parts of the Works.

The contract documents comprise the Project Drawings (listed in the Fourth Schedule), the Project Specification, a Contract Cost Plan (annexed to the Appendix Part 2), and the Articles, Conditions, Appendix and Schedules. (The fact that the work will be carried out on site by the Works Contractors is referred to in the Second Recital.)

Synopsis

1 Intentions

- The Management Contractor undertakes to cooperate with the professional team (1·4).
- Specific obligations are set out in clause 1·5 and the Third Schedule. These include preparation of Programmes, entering into Works Contracts, being responsible for the standards of the contract, and providing those site facilities and services listed in the Fifth Schedule. The Management Contractor is responsible for continuing supervision, and for ensuring that the project is carried through in an economical and expeditious manner. He is also to keep detailed records for the quantity surveyor to verify the Prime Costs (1·5).
- The Management Contractor is fully liable to the Employer for any breach of the contract, including those occasioned through breaches of Works Contracts (1·7).
- The Architect must supply the Management Contractor with further drawings and documents necessary to explain and amplify the Project Drawings and Project Specification (1·10).
- The contract documents will be Project Drawings, Project Specification, Articles, Conditions and Appendix, the Contract Cost Plan and the Schedules (1·3, also Articles 6 and 7).
- Quality is to be as described in the Project Specification, and in the Specification or bills of quantities for any Works Contract (3·8). There is no reference to Performance Specified Work in MC98, but there is in the Works Contract.
- The description of the Project is to be entered in the First Schedule.

JCT MC98

- Project Drawings are to be listed in the Fourth Schedule.

2 Time

- The Management Contractor proceeds to construction of the Project only after written notice by the Employer (2·1).
- At this point, the provisional dates (Appendix Part 1) are superseded by firm dates (Appendix Part 2) for possession and completion.
- The Management Contractor is given possession of the site on the date stated, and is required to secure commencement and ensure regular and diligent progress (2·3).
- Deferment of possession for up to six weeks is possible subject to an Appendix entry (2·3·2).
- Use or occupation of the site or the Project by the Employer prior to Practical Completion is provided for (2·3·4).
- Project extensions of time may be awarded to the Management Contractor by the Architect. The events or items which are relevant are few in number (2·13) and any extension requires the Management Contractor to have used its best endeavours to prevent delay (2·12).
- If the Management Contractor proposes to extend the contract period of a Works Contract, the Architect must first be notified and has the right to dissent (2·14).
- Completion of the Project may be on or before the completion date (2·3·1).
- Practical Completion of the Project is certified by the Architect and is subject to his or her opinion (2·4).
- If the Management Contractor fails to complete the Project by the completion date, the Architect issues a certificate of non-completion (2·9).
- The Employer's entitlement to liquidated damages depends on the issue of a certificate of non-completion (2·10).
- Defects to be made good after the Defects Liability Period must be scheduled by the Architect and delivered up no less than 14 days after its expiry (2·5). When the defects have been rectified, the Architect issues a certificate of completion of making good defects (2·6).
- The contract provides for partial possession by the Employer, subject to the consent of the Management Contractor (2·8).
- The contract provides for the issue of instructions to the Management Contractor to accelerate the work. It is possible to establish a completion date earlier than the date stated in Appendix Part 1, provided that clause 3·6 is shown to apply (3·6).

JCT MC98

3 Control

- The Management Contractor has to identify the management personnel employed on the Project and/or site in a list attached to the Second Schedule, and to name a site manager in the Appendix Part 2. The consent of the Architect is required to any changes (3·1 and 3·13).
- The Architect is to issue written instructions to the Management Contractor as are reasonably necessary (3·3). The instructions might require Project changes or Works Contracts variations (3·4). The Architect is also to issue instructions about provisional sums in Works Contracts (3·4).
- The Architect has the power to instruct the Management Contractor to postpone any work (3·5) and, an unusual term in JCT contracts, to accelerate work, including altering its sequence under stated circumstances (3·6).
- The Architect shall provide the Management Contractor with information on levels, setting out, etc. for the Project (3·7), and shall issue directions as necessary to the Clerk of Works (3·18).
- The Management Contractor must obtain vouchers to satisfy the Architect about the compliance of goods and materials (3·9). He must also comply with Architect's instructions concerning testing (3·10), removal from the site of work not in accordance with the contract (3·11), and the immediate making good of defective work (3·12).
- There is a bar to assignment of the contract without written consent (3·19). However, there is an option clause which if the Appendix Part 1 entry states that it is to apply, allows the Employer to transfer a right of action against the Management Contractor to persons with a subsequent interest in the completed Works (3·20).
- Items of work to be carried out by Works Contractors, and which are identified in the Contract Cost Plan or instructions, are subject to the Conditions in Section 8 of the Management Contract (8·1).
- In the control of Works Contracts, the Management Contractor's obligations in respect of the Employer and vice versa are fully set out (3·21). Alleged breaches by the Management Contractor or by Works Contractors are covered.
- The Management Contract allows for work not forming part of the contract to be carried out by persons directly engaged by the Employer whilst the Management Contractor still has possession of the site (3·23).

4 Money

- Payment is made by the Employer to the Management Contractor at agreed and stated intervals, on Interim Certificates issued by the Architect. These can be issued at both pre-construction and construction stages (4·1).

JCT MC98

- Amounts due to Works Contractors are included, and the Architect must direct the Management Contractor as to the amounts due, and may be required to inform the Works Contractor (8·3). There is an option of early final payment to Works Contractors, subject to certain safeguards (8·4).
- Payments to the Management Contractor include for the Prime Cost of the work actually carried out, reimbursement of amounts incurred by the Management Contractor's site staff and services, and an instalment of the management fee (4·5 and 4·6).
- Retention is not deducted from the instalment of the management fee, and is set at 3 per cent of the work not yet at Practical Completion stage. It is held in trust without obligation to invest, and unless the Employer is a local authority, the Management or Works Contractor may require it to be placed in a separate account. Half the retention is released at Practical Completion of the Project (4·7 and 4·8).
- Interim Certificates are issued by the Architect in accordance with the periods stated in the Appendix in respect of the pre-construction period, and the period for construction (4·2).
- Certificates must show the amount due to the Management Contractor, and the basis of calculation. The final date for payment by the Employer is 14 days from the date of issue. Not later than five days after issue the Employer is to give written notice specifying the amount of payment proposed, and if sums are to be withheld the Management Contractor must be given further written notice not later than five days before the final date for payment (4·3).
- Where the Employer fails to give proper notices or does not pay amounts properly due by the final dates for payment, the outstanding sums will attract interest at 5 per cent over base rate, and will give the Management Contractor a right to suspend work (4·3).
- An Interim Certificate shall be issued no less than 28 days before the Final Certificate, which includes any outstanding sums due to Works Contractors (4·11).
- The Management Contractor must provide the quantity surveyor with all information necessary for finally ascertaining the Prime Cost no later than six months after Practical Completion. The Construction Period Management Fee may be adjusted in accordance with the contract (4·9 and 4·10).
- The Architect issues the Final Certificate not later than two months from occurrence of the latest of the events listed in the contract (4·12).
- The requirement for written notices by the Employer, and the application of interest on amounts properly due and not paid by the final date of payment, also applies in the case of the Final Certificate. The final date for payment is 28 days from the date of issue.
- Applications in respect of loss and expense made by a Works Contractor must be

passed to the Architect by the Management Contractor. If the Architect forms the opinion that the application is valid and meets the conditions set out in the Works Contract, the quantity surveyor can be instructed to ascertain the amount in collaboration with the Management Contractor (8·5).

5 Statutory obligations

- The CDM Regulations require the appointment of a Planning Supervisor and a Principal Contractor. Various other obligations arise, particularly concerning the Health and Safety Plan and the Health and Safety File. Provision is made for these in the contract (5·18 to 5·21).
- Responsibility for compliance with legislation and serving notices rests with the Management Contractor (5·1). If the Management Contractor finds a divergence between statutory requirements and contract documents or further drawings etc. he is to inform the Architect, who must issue instructions within seven days (5·2).
- The Management Contractor is empowered to take action in any emergency to ensure compliance, and subject to certain conditions, this will be deemed a variation to the Management Contract or a Works Contract as applicable (5·4).

6 Insurance

- The Management Contractor indemnifies the Employer in respect of personal injury or damage to property other than the actual Works (6·7 and 6·8). This is to be backed by insurance (6·10).
- If instructed, the Management Contractor is to take out joint names insurance for the Employer against the risk of legal nuisance. There is a list of exceptions, and damages must not be directly attributable to any negligence by the Management Contractor or Works Contractors. An Appendix entry including the amount of cover is required (6·11·1).
- Where clause 6·4A is to apply, insurance of 'the Project' (i.e. work executed and site materials) is to be taken out by the Management Contractor in joint names (i.e. in his and the Employer's name), for the full reinstatement value of the Project against all risks. This must be done before any work begins on site.
- Where clause 6·4B is to apply, the requirement to take out such insurance rests with the Employer.
- In the event that terrorism cover is withdrawn and is no longer available, the situation and options open to the Employer are dealt with in clauses 6·4·10 and 6·5·4, introduced by Amendment 4.
- Where clause 6·5 is to apply, and the Project comprises alterations of or an extension to existing structures, then the Employer is required to take out a joint names policy in respect of the existing structures and contents. This is to be for the full cost of

JCT MC98

reinstatement etc. in the event of loss due to Specified Perils.

- Insurance for the Employer's loss of liquidated damages is an option (6·6 and Appendix Part 1).
- An Appendix entry will show whether the Joint Code of Practice on the Protection from Fire of Construction Sites is to apply (6FC) and if so, both the Employer and the Management Contractor must comply with it.

7 Termination

- The Employer is allowed to determine the employment of the Management Contractor for reasons of default (7·2). A warning notice may be issued by the Architect, but the notice of determination is a matter for the Employer.
- In the case of insolvency of the Management Contractor, and depending on the circumstances, determination might be automatic subject to possible reinstatement, or the Employer might elect to enter into an agreement (a '7·5·2·1 Agreement') to allow continuation or novation (7·5).
- The Employer is allowed to determine the employment of the Management Contractor at will (7·20). This may of course happen either before or during the Construction Period.
- The Management Contractor is allowed to determine his own employment for reasons of default by the Employer (7·9).
- Either party can determine the employment of the Management Contractor for listed neutral causes (7·13).
- The respective rights and duties of the parties concerning payment, removal and completion are set out in detail (7·6, 7·11 or 7·14).

8 Miscellaneous

- A list of definitions is included (1·3).
- Access for the professional team is assured, but subject to restrictions to protect any proprietary rights of the Management Contractor and Works Contractors (3·17).
- The Architect may order the removal of the manager from the Project and the Management Contractor must find a suitable replacement (3·14) subject to approval by the Architect.
- Where progress is disturbed because of the discovery of antiquities, the Management Contractor is obliged to inform the Architect, who must issue instructions (3·26 and 3·27).
- There is a contracting out of third party rights under the Contracts (Rights of Third Parties) Act 1999, introduced by Amendment 2.

JCT MC98

- Section 8 also deals with relevant issues arising from the Works Contracts, in particular the terms of such contracts, nominated suppliers to Works Contractors, duties of the Management Contractor under Works Contracts, final payment and loss and expense reimbursement to Works Contractors.
- The five Schedules which form part of MC98 are very important. They are:

  First Schedule: Project description: a short statement of the scope of the project, to be completed by Employer;

  Second Schedule: definition of Prime Cost payable to the Management Contractor relating to Works Contracts; on-site staff of Management Contractor; on-site labour, materials, goods, plant, stores and services provided by Management Contractor;

  Third Schedule: services to be provided by the Management Contractor: can be selected from a list of more than 50 obligations relating to both the pre-construction period and after work starts on site;

  Fourth Schedule: list of project drawings, all to be signed by the Management Contractor and Employer;

  Fifth Schedule: site facilities and services to be provided by the Management Contractor: to be completed before the construction of the Project, and initialled at the same time as Appendix Part 2 is signed.

9 Disputes

- Part II of the Housing Grants, Construction and Regeneration Act 1996 gives either party a statutory right to refer any difference or dispute arising out of the contract to adjudication. Article 8 provides for this.
- Procedures for referral to adjudication, the appointment of an adjudicator, the powers and conduct of an adjudicator are as set out (9A).
- Article 10A establishes arbitration as the agreed method for final determination of disputes (9B) unless the Appendix entry shows that this has been deleted in favour of legal proceedings (Article 10B and 9C).

JCT MC98

This contract?

If considering using MC98 remember that:

It is intended for use where the Employer has appointed a Contract Administrator, a quantity surveyor and other advisers to make up a professional team, and the team has prepared project drawings and a project Specification, and later detailed drawings, Specifications and bills of quantities for works packages. The Works Contractors enter into contracts direct with the Management Contractor. JCT MC98 is not a lump sum contract, and the sum paid by the Employer to the Management Contractor is the Prime Cost of the work together with a management fee.

The management contract is in one version only, for use in the private or public sectors. The Conditions apply to both the pre-construction period, and construction period and the operative details are entered in Appendix Part 1 and Appendix Part 2 respectively. The Conditions include deferment of possession, acceleration, partial possession, performance specified work but not Contractor's design. There is also a section relating to the Works Contractors and respective obligations.

When completing the form decisions are required relating to deferment of possession; insurance of the Project; liquidated damages; acceleration; management fee; Joint Fire Code; and EDI. The five Schedules should also be checked for completeness of entries.

If acting as Contract Administrator it means dealing directly with the Management Contractor, who in turn will be involved in the administration of Conditions in the Works Contracts. There are nevertheless some instances when the Contract Administrator will be involved with the Works documents, and the procedural rules, which can become complicated, need to be meticulously observed. The whole process demands constant monitoring and coordination if unnecessarily abortive work is to be avoided. The resulting documentation can be voluminous with this type of contract, and consequently quite arduous in terms of administration.

When the predecessor to MC98 first appeared it was the first such standard form available. Since then this procurement method seems to have declined in popularity, giving way to construction management as developer clients become more sophisticated with impressive in-house expertise. It is a relatively high risk contract with imprecise cost and time elements initially. It depends on goodwill and a high degree of trust between the Employer, the professional team, and the Management Contractor.

JCT MC98

Related matters

Documents

Standard Form of Management Contract
1998 Edition (MC98)
Amendment 1: 1999 (Construction Industry Scheme)
Amendment 2: 2000 (sundry amendments)
Amendment 3: 2001 (terrorism cover/Joint Fire Code)
Amendment 4: 2002 (extension of time)
Phased Completion Supplement for Management Contract

References

JCT Practice Note MC/1: Management Contracts under the JCT Documentation (1987)
JCT Commentaries on the JCT Management Contract Documentation MC/2 (1987)
(both written for MC87 but still largely relevant to MC98)

Commentaries

Vincent Powell-Smith and John Sims
The JCT Management Contract: A Practical Guide
Kluwer Publishing (1988)

JCT MC98 WORKS

The Joint Contracts Tribunal Ltd

Works Contract 1 and 2 and 3 1998 Editions for use with Standard Form of Management Contract

Background

Contract Documentation for use with the JCT Standard Form of Management Contract includes the following:

- Works Contract/1 (in three sections each of which is in part format, namely Section 1 Invitation to Tender; Section 2 The Tender; Section 3 Articles of Agreement)
- Works Contract/2 (Works Contract Conditions)
- Works Contract/3 (Employer/Works Contractor Agreement)

They can be used only with the 1998 Edition of the JCT Management Contract, and although this is largely a matter for the Management Contractor and each Works Contractor, effective administration of the main Management Contract often necessitates reference to the Conditions in the Works Contract.

Nature

The Articles of Agreement (Works Contract/1 Section 3) contain Recitals and Articles. The agreement is between the Management Contractor and the Works Contractor, and the First Recital confirms that the Management Contractor has entered into the JCT Standard Form of Management Contract.

The Recitals refer to 'numbered documents' which are effectively contract documents. In the event of any conflict between documents, the Works Contract Conditions are to prevail over numbered documents (1·4). If bills of quantities form part of the Works Contract documents, they must be to SMM7 unless stated otherwise (1·10).

Article 1 (in Works Contract/1 Section 3) establishes Works Contractor's obligations. Article 2 refers to a Works Contract Sum which will be a lump sum, and is VAT exclusive.

The agreement can be executed under hand but not as a deed, or as a deed. However, if the Management Contract is a deed then the Works Contract should be similarly executed.

Use

The choice of Works Contractors is a matter for agreement between the Architect and the Management Contractor, although the agreement is of course between the latter and the selected Works Contractor.

JCT MC98 WORKS

The Architect has relatively few direct obligations to Works Contractors. The Architect's opinion is the overriding factor concerning Practical Completion; the value of off-site goods which may be included in amounts due under a certificate and in respect of which a direction for payment is issued by the Architect. Suppliers may be nominated by the Architect. Administration of the Works Contract is mostly a matter for the Management Contractor, although the Architect must be notified on certain matters (e.g. extensions of time, compliance with directions, determination, etc.).

Synopsis

(clause numbers refer to Conditions of the Works Contract)

1 Intentions

- The Works Contractor is obliged to carry out and complete the Works in accordance with the Works Contract (1·7). This clause also refers to standards, and obligations relating to design and performance. The express reference to a design obligation is a reminder of the importance of Works Contract/3 which is a warranty to the Employer relating to design.
- The Works Contractor is to indemnify the Management Contractor for liability to the Employer in the event of defaults in performance under the Works Contract (1·8).

2 Time

- The time schedule for carrying out the work will have been established in Works Contract/1 Section 2. Starting will depend to a large extent on general progress of the Project, and the Management Contractor is obliged to keep the Works Contractor sufficiently informed before notice to commence is given (2·1).
- The Works Contractor must give written notice to the Management Contractor in the event of delay, and also state the causes. Subject to receipt of notice, particulars and an estimate of delay, the question of granting an extension rests with the Management Contractor. However, he must first notify the Architect, who has a right to dissent from the proposed action (2·2 to 2·10).
- If the Works Contractor fails to complete on time, the Management Contractor may be entitled to recover payment from the Works Contractor for any direct loss and expense (2·11 and 2·12).
- Practical completion is largely a matter for the Management Contractor and Works Contractor, but the consent of the Architect is needed before the Management Contractor can issue a Certificate of Practical Completion of the Works (2·13 and 2·14).

3 Control

- The Management Contractor is obliged to issue further instructions, directions, drawings, Schedules as necessary for carrying out and completing work under the Works Contract (3·1).

JCT MC98 WORKS

- Relevant instructions issued by the Architect under the Management Contract must be passed down, and the Management Contractor is then empowered to issue directions to the Works Contractor (3·3 and 3·4).
- The Works Contractor may make reasonable objection to a preliminary instruction issued under the Management Contract to accelerate work or alter its sequence (3·4). The Architect must take this into account, but if the notice stands, then the Works Contractor must comply. However, he can expect to receive payment for this (3·4).

4 Money

- Statutory obligations concerning payments apply following the introduction of the Housing Grants, Construction and Regeneration Act 1996 (Part II).
- The Works Contractor is paid by the Management Contractor as directed by the Architect in Interim Certificates (4·18). These sums are subject to retention. The valuation includes for work properly executed, unfixed goods and materials properly on site, and it may also include 'listed' off-site materials and goods (4·22).
- Retention in respect of Works Contracts is reduced when Practical Completion of the Works Contract (not of the Project under the Management Contract) has been reached (4·23).
- The Works Contractor must provide the Management Contractor, and if instructed the quantity surveyor also, with all information necessary for computation of the Ascertained Final Works Contract Sum (4·30) (there is no Final Certificate relating to Works Contracts, only that issued under clause 4·12 of the Management Contract).

5 Statutory obligations

- The CDM Regulations require the appointment of a Planning Supervisor and a Principal Contractor, but this will normally be covered in the main contract MC98. Statutory obligations which are also likely to implicate Works Contractors will include those relating to the Health and Safety Plan and the Health and Safety File (5E).
- The express obligations on the Works Contractor are concerned with tax matters to a large extent.

6 Insurance

- The Works Contractor indemnifies the Management Contractor in respect of personal injury or death, and damage to property (which may include the Project), and is required to insure against this (6·4 and 6·5).
- Concerning loss or damage to the Project and to the Works, Works Contractors are entitled to the benefits of the joint names policy for the Project (i.e. 6·6 to 6·8), but this is not in respect of all risks, only for specified perils (defined in 1·3).

JCT MC98 WORKS

7 Termination

- The Management Contractor has the right to determine the employment of the Works Contractor for reasons of default. Determination upon insolvency is automatic (7·1).
- The Works Contractor has the right to determine his own employment for reasons of default by the Management Contractor (7·7).
- Rights and obligations following determination are clearly stated (7·4 and 7·5) .

8 Miscellaneous

- The provisions for nominated suppliers follow those found in JCT98, and usually are covered by a Prime Cost or provisional sum included in the numbered documents. Nomination is a matter for the Architect (8·1).

9 Disputes

- Statute requires an adjudication provision in all construction contracts. Both parties have the right to refer any dispute or difference arising out of the contract to adjudication. The decision of the adjudicator will be binding, at least until a final determination of the dispute by arbitration or legal proceedings (Article 3 and 9A).
- Article 4A confirms arbitration as the agreed method for final resolution of disputes (9B), unless this is deleted in favour of legal proceedings (Article 9B and 9C).

This contract?

If considering using WC98 remember that:

It is intended solely for use with the JCT Management Contract MC98, and indeed must be used with that contract.

Tendering is a matter for the Management Contractor, but the invitation might need to be checked for compatibility with the Management Contract Conditions. Similarly the tender is a matter between the Management Contractor and Works Contractor. Selection of the Works Contractors is by agreement between the Management Contractor and the Contract Administrator. The latter has a particular role in certifying amounts due to the Works Contractors, applications for loss and expense, and the certifying of Practical Completion.

The Works Contract Conditions are section headed and synchronise with those in the Management Contract. If MC98 is used, then there is no choice other than to use WC98.

12 Management procurement: management forms

JCT MC98 WORKS

Related matters

Documents

Works Contract/1 (1998)
Section 1 Management Contractor's Invitation to Tender
Section 2 Works Contractor's Tender
Section 3 Articles of Agreement: Attestation

Works Contract/2 (1998) Works Contract Conditions
Amendment 1: 1999 (Construction Industry Scheme)
Amendment 2: 2000 (sundry amendments)
Amendment 3: 2001 (terrorism cover/Joint Fire Code/CIS/SMM)
Amendment 4: 2002 (extension of time/loss and expense)

Works Contract/3 (1998) Employer/Works Contractor Agreement

Phased Completion Supplement for Works Contracts

Construction management

The Joint Contracts Tribunal Ltd

Construction Management Agreement C/CM

Background

The fundamental distinction between management contracts and construction management lies in the degree to which the Client accepts a direct contractual relationship with the Contractors who carry out the work packages. With management contracting this will be achieved though the Management Contractor, and there will also be an independent team of professionals, including an architect and quantity surveyor with overall responsibility for design and contract administration.

With construction management, the Construction Manager will be the key person or firm with an overall responsibility for coordination and contract administration relating to the Trade Contracts. The Client also has a significant contribution to make, and is responsible for the engagement of a Consultant Team and nomination of a Consultant Team Leader. The Consultant Team will have a major involvement in the pre-contract period, and although likely to have some involvement during the construction period, this will not be in contract administration.

In July 1995 JCT produced draft documentation for construction management. This was referred to the constituent bodies of the JCT, but was then overtaken by the publication of the Latham Report, and the Housing Grants, Construction and Regeneration Act 1996 (Part II). The draft was developed to take account of the full implications of these, and the construction management documentation eventually appeared in 2002.

Nature

The documentation consists of an Agreement (C/CM) between the Client and the Construction Manager, and a Trade Contract (TC/C) between the Client and each of the Trade Contractors. There is also an Invitation to Tender, a Tender document, and warranties by a Trade Contractor to a purchaser or tenant, and funding organisation for the Project.

The Agreement (C/CM) between the Client and the Construction Manager is an attractively presented document, logically structured and with particularly clear layout making referencing relatively straightforward.

There are four Recitals, the first of which refers to the building works being phased, and a deletion is needed if this is to be a single stage operation. The Client undertakes to appoint the Consultant Team, and to have prepared an Initial Brief and preliminary

Construction management

Project Cost Plan. The other Recitals refer to the fact that the full CDM Regulations will apply; the services to be performed by the Construction Manager; and that the work will be carried out under contracts directly between the Client and Trade Contractors.

There are five Articles, which refer briefly to the obligations of the Construction Manager and the Client; the identity of the Planning Supervisor and Principal Contractor; and the methods for resolving disputes.

The Conditions are relatively short (just 19 pages in total) and section headed as follows:

1. Intentions of the Parties
2. Obligations of the Construction Manager
3. Obligations of the Client
4. Assignment and Sub-contracting
5. Payment
6. Insurance and Indemnities
7. Termination of Engagement of Construction Manager
8. Proper Law and Disputes

Probably the most significant parts of the document are the Schedules (taking up 35 pages in total). These act as reference points for much of what is carried in the Conditions, and are:

- First, Description of the Project;
- Second, Insurance and Indemnities;
- Third, Definition of Reimbursable Cost;
- Fourth, Model Services to be provided by the Construction Manager;
- Fifth, Construction Manager's Personnel;
- Sixth, Site Facilities and Services to be provided by the Construction Manager;
- Seventh, Construction Management Fee;
- Eighth, Consultant Team;
- Ninth, Cost Planning and Control;
- Tenth, Adjudication;
- Eleventh, Arbitration.

Construction management

Some of the Schedules contain essential detailed information, and others call for entries to be made. They amplify many of the provisions found in the Conditions, and are at the very heart of the agreement.

Perhaps surprisingly for a form more suited to major projects, attestation may be under hand or as a deed.

The nature of construction management procurement makes it likely to be of interest only for major building works undertaken by experienced client bodies. Although the Construction Manager is largely responsible for the management and coordination of Trade Contractors, this still leaves the client with considerable executive responsibility on a day-to-day basis for the duration of the pre-construction and construction stages of the Project.

Use

Any role for architects will be as members of the Consultant Team to which they might be appointed under a 'Consultancy Agreement'. They might also be named as Consultant Team Leader particularly for design stages at pre-construction, but much will depend on the nature of the work.

In the Eighth Schedule Part A the Client is to name the Team Leader, and to summarise the scope of the work of each member. A copy of the Client's Initial Brief and the Client's Preliminary Cost Plan will be developed into the Project Brief and the Project Cost Plan. The client must also identify the 'Client's Representative' and the 'Cost Consultant'. The measure of consultation between Construction Manager and the Consultant Team Leader is set out in commendable detail in Part C with clauses conveniently referenced. The involvement of the Consultant Team during the Pre-Construction and Construction Periods, which should be covered in any consultancy agreement, is clearly described in Part D of the Eighth Schedule.

Brief synopsis of Conditions

(although consultants are not involved directly in contract administration)

1 Intentions

- Subject to anything to the contrary the work will be carried out in phases (First Recital).
- Possession and control of the site is given to the Construction Manager, but this will not necessarily be exclusive to the Construction Manager (1·5).
- There is provision for the Client to use or occupy the site or the Project before completion (1·6).
- There is provision for partial possession by the Client (1·7).
- The Construction Manager is to manage the Project in accordance with the Project Brief, Project Cost Plan and Health and Safety Plan (2·1).

Construction management

- The Construction Manager must provide the services set out in Fourth Schedule Part A (Pre-Construction) and Part B (Construction) (2·1), and site facilities (Sixth Schedule).
- The Construction Manager must exercise skill, care and diligence, to the extent expected of a reasonably competent Construction Manager, and must carry professional indemnity insurance as indicated in the Appendix (2·5).
- The Construction Manager is not liable to the Client for the design of the Project (Eighth Schedule item 2·6).
- The Client may appoint a Client's Representative to handle all functions ascribed to the client (1·4).
- The Client will appoint the Consultant Team (named in First Recital), name the Leader (3·2) and may appoint a Cost Consultant (3·5).
- The Client will appoint Trade Contractors, taking into account the views of the Consultant Team. Appointments will normally be made after interview, an analysis of the tenders, and a written report by the Construction Manager (Fourth Schedule Part A).
- Contract documents comprise the completed Recitals, Articles, Conditions, Schedules and Appendix to the Agreement (C/CM). Documents referred to include the Initial Brief, Preliminary Cost Plan, and developed Project Brief, Project drawings, Project Programme, Project Specification, and Project Cost Plan (Eighth Schedule Part D item 1·2).
- Further drawings, Specifications, details and Schedules will be provided as necessary to explain or amplify the Project information (3·8).

2 Time

- The agreement refers to the work as relating to a pre-construction period and a construction period.
- There is no start and completion date with this agreement as found with traditional forms, and the construction period runs from the date of commencement of work on site by the Trade Contractor first on the scene, and ends with the date of issue of the Interim Project Completion Certificate (1·3).
- There will be a Project Programme prepared by the Construction Manager after consultation with the Consultant Team. This will identify critical path, lead times, and key milestones (Fourth Schedule item 5·1).The Construction Manager will update and expand this programme in liaison with the Client and Consultant Team (Fourth Schedule item 5·4).
- The Construction Manager will, before Trade Contract tenders are invited, prepare a detailed week by week programme (Fourth Schedule item 5·2).

Construction management

- During construction the Construction Manager will advise the Client on matters which may cause delay in completing the Project (Fourth Schedule item 9·8), and report on all matters related to progress after consulting the Consultant Team (Fourth Schedule item 9·11).
- The Construction Manager is not to be held responsible for delay, where this has been caused by members of the Consultant Team, but is required to use 'all reasonable efforts' to avoid or mitigate the effects (3·3).
- The Construction Manager, with agreement by the Consultant Team Leader, will certify Practical Completion in respect of each Trade Contract (2·2).
- After completion of the last Trade Contract, the Construction Manager with agreement by the Consultant Team Leader, will issue the Interim Project Completion Certificate and the Defects Liability Period (normally six months from the day named in the Interim Project Completion Certificate) will commence (2·2).
- When defects in work carried out under all Trade Contracts have been made good, the Construction Manager will issue a certificate of making good defects in respect of the Project (2·3).
- When the Construction Manager advises the Client that all obligations under the Trade Contracts have been fulfilled, the Client will issue a Final Project Completion Certificate (2·4).

3 Control

- Names of the Construction Manager's personnel are to be entered in the Fifth Schedule and any changes require the consent of the Client (2·2).
- Neither party to the agreement may assign the agreement without written consent (4·1).
- The Construction Manager cannot sub-contract his obligations without the Client's written consent (4·2).
- The Client will appoint the Trade Contractors, after receiving recommendations from the Construction Manager and where relevant the Consultant Team Members (3·6·1).
- The Trade Contracts will be on the current JCT Trade Contract (TC/C) unamended, unless otherwise agreed with the Client (3·6·2).
- The Client may issue to the Construction Manager such instructions in writing as are reasonably necessary (3·7).
- The Construction Manager will manage and coordinate the work of Trade Contractors (Fourth Schedule item 9·2).
- The Construction Manager will hold regular meetings with Trade Contractors and provide monthly written reports for the Client and Consultant Team Leader (9·4).

Construction management

- The Construction Manager will arrange and chair regular site progress meetings, to which the Client and the Consultant Team will be invited. He is also responsible for the minutes (9·5).

4 Money

- The Client undertakes to pay to the Construction Manager a Pre-Construction Period Management Fee as entered in Part 1 of the Seventh Schedule. This sum, exclusive of VAT, can be adjusted if circumstances change (5·8).
- Payment will be due in part on signing the agreement and thereafter at agreed intervals to be stated in the Appendix.
- In the construction period, the Construction Manager will submit to the Client monthly accounts which show sums due in respect of the Construction Period Management Fee (as shown in the Seventh Schedule); the reimbursable cost (as defined in the Third Schedule); and any other sums due to the Construction Manager (5·3).
- The usual five days' notices are required from the Client (5·5) and the final date for payment is 17 days from the date of issue of the Construction Manager's account (5·5).
- Failure to pay will attract interest at 5 per cent over base rate, and could result in the Construction Manager suspending work until payment is made (5·5).

Final payment to the Construction Manager is dependent on the final account showing the total sum due under the agreement. This should be submitted not later than one month after the date of issue of the Final Project Completion Certificate (5·6).

The usual five days' notices apply, and the final date for payment is 17 days from the date of issue of the Construction Manager's account (5·7).

5 Statutory obligations

- The CDM Regulations apply in full (Second Recital).
- Where the Construction Manager is the Principal Contractor he must ensure that the Health and Safety Plan complies with the Regulations (2·8).
- Where the Construction Manager is appointed Planning Supervisor he must ensure compliance with the CDM Regulations, and in particular Regulations 14 and 15 (2·9).
- The Construction Manager is to liaise with statutory authorities and statutory undertakers relating to site services (4·1).
- The Construction Manager is to advise the Client on orders to be placed with statutory bodies, and to manage implementation of their work (4·2).
- The Construction Manager will monitor Trade Contractors' compliance with statutory requirements and verify that all necessary approvals have been obtained (4·3).

Construction management

6 Insurance

- Insurance and indemnities are dealt with in the Conditions by a very brief reference to the Second Schedule.
- The Construction Manager is to indemnify the Client against claims relating to personal injury and death, and against damage to property real or personal, other than the Project (Second Schedule item 13).
- The Construction Manager must take out insurance in respect of claims for personal injury or damage to property for the sum entered in the Appendix (Second Schedule item 17·2).
- The liability of the Construction Manager under this agreement is limited to the sum entered in the Appendix, provided that the Appendix states that clause 2·7 is to apply.
- The Construction Manager may be required to take out professional indemnity insurance for an amount entered in the Appendix (2·5).
- Insurance of the Project will be taken out by the client under a joint names policy for all risks cover, for the full reinstatement value of the Project and replacement value of site facilities, plus the amount entered in the Appendix for professional fees. Where the Project comprises work to existing structures, cover must include the contents owned by the Client, but this will be only in respect of Specified Perils (Second Schedule item 1).

7 Termination

- Termination relates to the Construction Manager's engagement.
- Termination may be by the Client in the event of the Construction Manager's insolvency, or failure to exercise the degree of skill, care and diligence required under the Agreement (7·2).
- The Client may also terminate the engagement of the Construction Manager at will (7·3).
- The Construction Manager may terminate his own engagement in the event of the client's insolvency; failure to pay amounts properly due; or if work is suspended for a continuous period of six months (7·4).
- The consequences of termination are set out in detail (7·6).

8 Miscellaneous

- Third party rights under the Contracts (Rights of Third Parties) Act 1999 are excluded (1·12).
- The Agreement is to be construed in accordance with the law of England (8·1).

Construction management

- The Agreement is not for use under Scots law.

9 Disputes

- Article 5 confirms the right of either party to seek adjudication in the event of a difference or dispute (Article 5).
- Nominators for an adjudicator can be agreed and shown by appropriate deletions made in the Appendix (Tenth Schedule item A·2).
- The conduct of the adjudication and effects of adjudicator's decision are covered in the Tenth Schedule.
- Final determination will be in the English courts (does this exclude Wales?), unless the Appendix shows that clause 8·3 (arbitration) is to apply.
- Where arbitration applies, appropriate deletions in the Appendix will indicate the appointors of an arbitrator. The conduct of the arbitration and effects of the award are covered in the Eleventh Schedule.
- A footnote to Section 8 of the Conditions is a reminder that disputes may also be resolved by the process of mediation.

Construction management

This contract?

If considering using C/CM 2002 remember that:

It is intended for use with large projects where the Client wishes to enter into separate contracts with members of the Consultant Team who will be responsible for design, the Construction Manager who will provide services during both the pre-construction period and the construction period, and each of the Trade Contractors who will carry out and complete the Works. The Client assumes the central role, although a Client's Representative and a Cost Consultant may also be appointed. This is not a lump sum agreement (although the Trade Contracts may be), and the Client will pay the Construction Manager reimbursable costs and a management fee.

The form is not for use in Scotland, and the SBCC have decided not to publish a Scottish version.

The agreement is in one version only, and is the head contract in a standard construction management pack of documents. The Conditions apply to both the pre-construction and construction periods, and the operative details are entered in a single Appendix. The Conditions include for completion in phases and partial possession.

When completing the form decisions are required relating to preparing the Project Cost Plan; insurance cover including professional indemnity insurance; limitation of liability; any amendments to the standard Trade Contracts; reimbursable costs; and payments to the Construction Manager during the pre-construction period. The 10 Schedules are a particularly important part of the Agreement and should be checked for content and completeness of entries.

If acting as design consultant, remember that the Construction Manager may also advise the Client in preparing the Project Brief, make recommendations and review design and other drawings. A close working relationship with the Consultant Team is essential.

If acting as Client's Representative, then remember that this could involve carrying out all the functions ascribed to the Client, acting as agent unless the Agreement specifically states otherwise.

Clearly this is an arrangement which will appeal only to an experienced client probably with in-house services. It is relatively low risk for the Construction Manager. The Agreement has the merit of being logically structured, clearly laid out, and in taking only 64 pages to cover a very sophisticated operation.

Construction management

Related matters

Documents

Construction Management Agreement C/CM 2002
Trade Contract TC/C

Tender Document TC/T
Part 1: Invitation to Tender
Part 2: Tender by the Trade Contractor
Warranty TCWa/P&T (warranty to purchaser or tenant)
Warranty TCWa/F (warranty to funder)
Fluctuations Code

References

Earlier JCT Practice Notes:
Practice Note No 27: Application of CDM Regulations (1995)
Practice Note No 28: Mediation (1995)

Series 2 JCT Practice Notes (yellow covers):
Practice Note 1: Construction Industry Scheme
Practice Note 2: Adjudication (includes text of agreements)
Practice Note 3: Insurance, Terrorism Cover
Practice Note 4: Partnering (includes text of non-binding charter)
Practice Note 5: Deciding on the Appropriate JCT Form of Contract
Practice Note 6: Main Contract Tendering (includes model forms)

JCT Guide to Construction Management Documentation

Construction management

The Joint Contracts Tribunal Ltd

Trade Contract TC/C

Background

This is the dedicated contract to be used between the Client and each Trade Contractor (3·6·2 of C/CM Agreement). However, the clause also accepts that a Special Trade Contract may be used in lieu by agreement, or amendments to TC/C may be required by the Client.

Nature

Where acts or decisions are expressly ascribed to the Client (or if applicable the Client's Representative), then references mean precisely that. The main channel for contract administration, however, will be the Construction Manager who is to be regarded as the agent of the Client. There is no direct contractual relationship between a Trade Contractor and the Consultant Team or the Consultant Team Leader, but consultants need to have a reasonable working knowledge and understanding of the Trade Contract in order to appreciate and anticipate the degree of support and cooperation on which the Construction Manager's actions may depend.

This contract is relatively conventional in format and structure. The Agreement, which may be executed under hand or as a deed to enable compatibility with C/CM, includes eight Recitals and seven Articles. The First Recital refers to the use of the standard Tender Document TC/T, completed copies of which are to be annexed to the Articles. The name of the Consultant Team Leader is given in the Tender TC/T Part 1.

The Conditions run over 80 plus pages (compared with 19 for the C/CM Agreement), and are section headed. These include uniquely specified suppliers; performance specified work; and fluctuations (the detailed provisions of which are published separately). There are four Schedules relating to a bond in lieu of retention; collateral warranties (which if used are to be on the JCT Trade Contractor Warranty forms); Performance Bond (not a JCT document); Parent Company Guarantee (again not a JCT issue). The Appendix is followed by Annexes in respect of EDI, Advance Payment Bond, bond in respect of payment for off-site materials, and Formula Adjustment.

Use

Except for the Consultant Team Leader being co-signatory to Interim Certificates, members of the Consultant Team have no express role under this contract. Administration is in the hands of the Client or the Construction Manager. Architects may of course act as the Construction Manager, and nearly all consultants will need to work in close collaboration with the Construction Manager during the course of the Works.

Construction management

Brief synopsis of Conditions

(likely to be of interest to the Consultant Team)

1 Intentions

- Trade Contract tender documents may include drawings, together with a Specification, bills of quantities, or Schedules of Work. Any of the last three may be a 'Priced Document' perhaps together with a Trade Contract Sum Analysis or a Schedule of Rates on which the Contract Sum is based.
- The Trade Contractor may be required to provide design for the Works (1·9).
- The Trade Contract documents include the Articles, Conditions, Appendix and completed Schedules. In the event of any conflict, nothing shall override or modify the printed text (1·11).
- Further necessary drawings and Schedules may be issued by the Construction Manager to amplify or explain the Trade Contract Documents (1·18).
- Drawings, details etc. must be returned to the Construction Manager if requested (1·20).
- Performance Specified Work may be included, and 'as-built' information required (1·22).

2 Time

- A commencement date and a completion period are entered in the Appendix (2·1).
- The Trade Contractor must work in accordance with the progress of the Project (2·1) and occupation of the site will be as permitted by the Construction Manager.
- The completion period may be revised in the event of delay by a Relevant Event (2·3).
- The Construction Manager may issue instructions to postpone any work (2·1).
- The Construction Manager may require the Trade Contractor to submit a written quotation for accelerating the work (2·6).
- Failure by the Trade Contractor to complete to time may result in payment of direct loss and/or expense to the client (2·11). There will be a limit to financial liability if a sum is entered in the Appendix (2·12).
- Practical Completion of the Works is certified by the Construction Manager (2·13).
- Defects liability runs from Practical Completion to the end of the Project Defects Liability Period, which normally runs for six months after Interim Project Completion is certified (2·16).
- Phased completion (2·20) and partial possession by the Client are possible (2·21).

Construction management

3 Control

- The Trade Contractor must keep a person-in-charge continually on site (3·1).
- Instructions are issued by the Construction Manager and must be in writing (3·4).
- The Construction Manager is responsible for levels and providing setting out information (3·6).
- Access to the Works for the Client, Construction Manager and any other authorised person is provided for. Presumably this could include Consultant Team Members (3·7).
- The Construction Manager may issue a Variation Instruction, and the Trade Contractor may wish to submit a '3A Quotation' to cover the costs, likely time and disruption factors, and need for additional resources (3·19 to 3A·8).

4 Money

- Payment by the client may be on the basis of either a Trade Contract Sum (a lump sum) or the Ascertained Final Trade Contract Sum.
- Where the Trade Contract Sum applies (Article 2·1) then valuation of variations can be by Alternative A – Trade Contractor's Price Statement, or Alternative B – valuation made by the Construction Manager using the valuation rules (4·4). Where Article 2·2 applies, the valuation of work comprised in the Works is by remeasurement based on rates or prices for measured work, or a Schedule of daywork prices (4·7).
- The Trade Contractor may be entitled to reimbursement of direct loss and expense due to matters listed (4·22).
- The Trade Contractor may make written application to the Construction Manager for payment either at intervals entered in the Appendix, or at agreed stages. Whether or not application is made the Construction Manager, not later than 14 days after the date or stage, must issue an Interim Payment Certificate signed jointly by the Consultant Team Leader and the Construction Manager (4·11).
- The usual notices by the Client apply, and final date for payment is 14 days from the date of issue of each certificate. Failure to pay may result in payment of 5 per cent interest over base rate, and the right of the Trade Contractor to suspend work (4·11).
- Retention will normally be at 5 per cent, and the Client acts in fiduciary as trustee but without obligation to invest. Alternatively, if so stated in the Appendix, a Trade Contractor's Bond may be used in lieu of retention (4·15).
- The Trade Contractor is to submit documents to the Construction Manager not later than three months after Practical Completion of the Works, in order that there can be final adjustments or ascertainment of the Final Contract Sum (4·17 and 4·18).
- Subject to conditions stated in the Trade Contract, the client is to issue a Final Project

Construction management

Completion Certificate and a Final Statement. The usual notices by the client apply, and the final date for payment is 28 days from the date of issue of the Final Statement (4·19).

5 Statutory obligations

- The Trade Contractor is to comply with statutory requirements and pay fees or charges not already allowed for in the Trade Contract documents. Otherwise the amounts will be added to the Contract Sum.
- The Construction Manager will issue instructions in the event of divergence from statutory requirements, or emergency compliance.
- The Trade Contractor is to provide the Principal Contractor with necessary information concerning the Health and Safety Plan, and the Planning Supervisor with required information for the Health and Safety File.

6 Insurance

- The Trade Contractor indemnifies the client in respect of claims arising due to personal injury or death (except where due to the client or any person for whom the client is responsible), and damage to property other than the Works provided that this is due to the negligence of the Trade Contractor (6·1 to 6·3). The Trade Contractor must take out insurance cover not less than the figure entered in the Appendix. This is the contractual limit, not necessarily the limit of liability (6·4).
- Insurance of the Project and of the Works is covered under Client and Construction Manager Agreement (C/CM), and the client is to ensure that Trade Contractors are to enjoy the benefits of the joint names policies and a waiver of subrogation rights (6·9).

7 Termination

- Termination relates to the employment of the Trade Contractor.
- Termination may be by the Client on the insolvency of the Trade Contractor, or in the event of default for reasons listed (7·2).
- Consequences of determination by the Client are set out in the contract (7·6).
- The Trade Contractor may determine his own employment if the Client defaults or becomes insolvent (7·9 and 7·10).
- Consequences of determination by the Trade Contractor are set out in detail (7·11).
- Either party may determine the employment of the Trade Contractor where there is no default but one or more of listed events occur (7·13) and the consequences of such determination are set out (7·14 to 7·18).
- The Client, except if he becomes insolvent, is entitled at any time to discontinue the Project and to require the Trade Contractor to cease work (7·19).

Construction management

- The contract provisions still leave other rights and remedies open.

8 Miscellaneous

- The contract is to be construed in accordance with the law of England (1·30).
- There are supplemental provisions for EDI subject to Appendix deletions.
- The provision for uniquely specified suppliers refers to suppliers named in the Trade Contract documents (8·1).
- Performance specified work requires each item to be identified (11·1).
- Third party rights under the Contracts (Rights of Third Parties) Act 1999 are excluded (1·32).

9 Disputes

- Article 6 confirms the right of either party to seek adjudication in the event of a difference or dispute (Article 6).
- Nominators for an adjudicator can be agreed and shown by appropriate deletions made in the Appendix.
- The conduct of the adjudication and effects of adjudicator's decision are covered (9A).
- Final determination will be at arbitration (9B) or by legal proceedings (9C) depending on an Appendix deletion.

This contract?

If considering using TC/C remember that:

It is intended solely for use with the JCT Client and Construction Manager Agreement (C/CM), and must be used with that document. The Trade Contract is not for use under Scots law.

Trade Contracts are a matter for the Client, but the necessary documents for tendering are supplied by the Consultant Team, and issued by the Construction Manager on the Client's behalf. Normally tendering will involve the TC/T Invitation and Tender documents, and lead to Trade Contract (TC/C) unamended.

The Trade Contract Conditions are a substantial 118 pages long, and section headed. The client may accept a Special Trade Contract or amendments to TC/C, but otherwise there is no choice but to use TC/C with the C/CM Agreement.

Construction management

Related matters

Documents

Standard Form of Trade Contract TC/C 2002 Edition
Tender Document TC/T
Part 1: Invitation to Tender
Part 2: Tender by Trade Contractor
Warranty TCWa/P&T (warranty to purchaser or tenant)
Warranty TCWa/F (warranty to funder)
Fluctuations clauses for use with Trade Contract TC/C

References

Earlier JCT Practice Notes:
Practice Note No 27: Application of CDM Regulations (1995)
Practice Note No 28: Mediation (1995)

Series 2 JCT Practice Notes (yellow covers):
Practice Note 1: Construction Industry Scheme
Practice Note 2: Adjudication (includes text of agreements)
Practice Note 3: Insurance, Terrorism Cover
Practice Note 6: Main Contract Tendering (includes model forms)

JCT Guide to Construction Management Documentation

Partnering arrangements

13

Partnering agreements

The Joint Contracts Tribunal Ltd
JCT Non-Binding Partnering Charter for Single Project 2001 Edition

The Institution of Civil Engineers (ICE)
NEC Partnering Option X12 First Edition (2001)

The Association of Consultant Architects Ltd
ACA Standard Form of Contract for Project Partnering PPC2000

'Partnering is neither a particular procurement approach, nor is it a particular type of contract: it is about culture and the way in which the participants view and manage the project' (JCT Note on Partnering).

Partnering as an ethos which attempts to avoid adversarial conflict was the subject of a number of studies and publications in the 1990s, and endorsed in both the Latham and Egan Reports. Arrangements may take the form of project partnering applicable to a single project, or strategic partnering which may embrace a number of projects over time.

Project Team Partnering is about working together to achieve the client's objectives for a project by adopting a management approach in which efficient coordinated working is measured against performance indicators and targets. There must first be some formalised expression of agreement between the partners, and this is usually achieved by using either a free-standing non-binding charter for a single project; a binding agreement for single project or strategic partnering, perhaps through the incorporating of additional clauses in a standard contract; or by using a form of contract specially drafted for multi-party partnering.

The JCT Ltd, ICE, and ACA have each taken different approaches to partnering, and it is understood that partnering using GC/Works forms is currently under review.

The Joint Contracts Tribunal Ltd

JCT Non-Binding Partnering Charter for Single Project 2001 Edition

Background

The Latham Report called for contract Conditions which included a specific duty for all parties to deal fairly, and that there should be firm duties of teamwork with a general presumption to achieve 'win-win' solutions to problems rather than apportion blame. Unlike some standard forms, most JCT contracts do not expressly call for parties to act with mutual cooperation and deal fairly, although this may well be implied.

But the climate is changing, and increasingly emphasis is placed on achieving improved, more efficient and more integrated team working through better management of the supply process. Hence the current interest in partnering.

This Charter is the first JCT document to refer specifically to partnering, and other perhaps more binding agreements may well follow. The Charter is for single project partnering, and although non-binding and free-standing, used in conjunction with JCT forms of contract the agreement of the signatories could well be considered material in any arbitration or court proceedings arising out of the building contract.

Nature

The form is a Model Charter, extremely simple and only three pages long. It is a formal document by which the team signatories agree to work together to complete a Project which meets the client needs, quality standards, cost, and Programme. The Team Members may include any involved in the Project, regardless of contractual relationships. In signing the Charter, the members agree to act in good faith, in an open and trusting manner, and in a cooperative spirit. They also agree not to put the blame on others, to respect each other's skills, and act fairly towards each other. It therefore determines what the conduct of the members should be.

Use

The Charter may be used to supplement any of the main JCT Forms of Building Contract. The purpose is to focus the minds of the signatories on the aims and mutual objectives of the Project, and how these can be better achieved through effective collaboration.

Although the Charter might seem disarmingly simple, using it to good effect is likely to mean considerable effort. The Charter lists, under four headings, objectives to be achieved. These obviously must be developed in detail, and also be capable of measurement against performance standards set by the Team. The process of partnering is likely to mean first, reaching agreement to engage in partnering between all involved; secondly, setting up a 'workshop' or focused discussion to agree objectives; and thirdly, maintaining a series of 'workshops' throughout the whole pre-construction and construction stages of the Project to monitor, adjust and control as necessary.

Synopsis

In this Charter there are no conditions, but the Conditions found in the main building contract and consultant appointing documents will need to be observed. Apart from being an overriding expression of intention, the Charter should be considered as a document bringing a commitment to a management process.

The Charter will need to be accompanied by partnering documentation. This should set out such things as composition of the Partnering Team, agreement on the Key Performance Indicators, targets and measurement arrangements, risk identification, arrangements for the operation of Partnering Team workshops and meetings, communications policy, information systems to be used in common etc.

This contract?

If considering using the JCT Partnering Charter remember that:

This is a non-binding agreement, and it is not a contract (although it could be an important factor in the construing of contract terms).

However, as a statement of intention by Team Members involved in a Project wishing to embrace the fashionable partnering ethos, this is a refreshingly simple and direct document. It can be used in conjunction with any JCT form of contract (or other form for that matter), but it seems particularly suitable for use with larger projects involving a number of key team players. The Charter of itself will not bring about partnering, it must be backed by relevant supporting documents and effective management procedures.

The JCT has stated its intention also to prepare an entirely new form of contract which includes partnering. Meantime the Charter should be an effective tool – partnering is very much a matter of attitude not words.

Related matters

Documents

JCT Non-Binding Partnering Charter for Single Project 2001

References

JCT Series 2 Practice Note 4: Partnering (contains a copy of the Charter)

The Institution of Civil Engineers

NEC Partnering Option X12 First Edition (2001)

Background

The Second Edition Engineering and Construction Contract requires the two parties, the Employer and the Contractor, to act as stated in the contract and in a spirit of mutual trust and cooperation. This main contract also provides for an early warning of matters likely to result in increased price, delayed completion or impaired performance of the Works in use.

This partnering option, which touches others beyond the main contracting parties, was published as a First Edition in June 2001. It can be used with any NEC contract except the Adjudicator's Contract. As the title implies, it is not a free-standing document, but is an option which may be incorporated into those NEC contracts for any Team Members involved with a Project, whether as Contractor, sub-contractors, consultants or sub-consultants. If the Option X12 is incorporated, then the parties to these contracts will have additional responsibilities in common.

Nature

The Option can be used for single project partnering or for strategic partnering over several projects. It can of course only be used with NEC contracts, and does not result in a multi-party contract.

The Option requires additional Contract Data, some of which will not change (e.g. Client's Objective and Partnering Information on agreed methods of operating), and a Schedule of Partners, and Schedule of Core Group Members, which might change during progress of the Works. The Schedule of Partners will include identity and contribution of the partners, joining and leaving dates, and details of Key Performance Indicators, targets, measurement arrangements and any incentive payments. The Schedule of Core Group Members will give identity of partners, and joining and leaving dates. Both Schedules will probably need revising from time to time.

Option X12 also includes 22 short clauses set out under four headings, and the document includes helpful guidance notes on these clauses.

Use

The Option is incorporated into the contracts of the partners by entering 'X12 (published by the ICE June 2001)' in the first line of Contract Data Part One, and also adding an entry 'Option X12' at the end of 'Optional Statements' in the Contract Data

Part One: Data provided by the Employer. The information for this entry is suggested in the text of the NEC Partnering Option, and covers details of the Client, the Client's Objective, and Partnering Information.

Synopsis of clauses

- The Option does not create legal partnerships outside contracts.
- Each partner collaborates to achieve the Client's Objective, and the stated objectives of every other partner.
- Each partner nominates a representative with authority to act.
- The Client is a partner.
- Partners are to cooperate over providing information, and giving early warnings of matters likely to affect other partners. Partners may give advice, information and opinion and if so it must be given fully, openly and objectively.
- A Core Group is selected by the partners, with authority to act on behalf of partners. The Core Group is led by the Client's Representative (a position not specifically defined in the Option or the contract).
- The Core Group is responsible for preparing a timetable of partner's contributions, and is pivotal to the partnering operation.
- Partnering Information is defined as that contained in the documents referred to in Contract Data, or in an instruction under the contract.
- The partners work together as stated in the Partnering Information in a spirit of mutual trust and use common information systems as set out in that Information.
- Partners will be paid the amount stated in the Schedule of Partners if the target stated for a Key Performance Indicator is improved upon or achieved.

This contract?

If considering using Option X12 remember that:

It is intended only for use with any NEC contract (except the Adjudicator's Contract). It brings more than the two main contracting parties into a partnering relationship, but it does not create a multi-party contract.

The responsibilities for Team Members included under Option X12 will be in addition to the contractual responsibilities which they might have.

There will be additional Contract Data, and additional clauses incorporated.

The Option is a neat way of bringing about partnering relationships, and the partnering clauses set out the actions necessary to achieve this.

Related matters

Documents

A new engineering contract document, the NEC Partnering Option: Option X12
First Edition (June 2001)

References

Guidance Notes and Guidance Notes on Clauses (included in document)
Construction Industry Council, Guide to Project Team Partnering

PPC2000

The Association of Consultant Architects Ltd

ACA Standard Form of Contract for Project Partnering PPC2000

Background

Partnering agreements and conventional contracts might be complementary, but there are essential differences between the function of a contract and a partnering arrangement. The former gives rise to contractual obligations strictly between the contracting parties, and ideally is couched in terms of certainty which, in the event of dispute, can be tested in the courts. Clear wording of the contract and adherence to sound administrative procedures are essentials. The latter seeks to establish collaborative working between all key players committed to putting the project first, invokes trust and fairness, and operates as a project management tool. Obviously there can be difficulties in trying to marry partnering principles to contract Conditions.

Flying in the face of this conventional wisdom, PPC2000 was produced by the Association of Consultant Architects Ltd, and drafted by David Mosey of Trowers & Hamlins, Solicitors. This is a brave attempt to bring together partnering arrangements, consultant appointment terms and a building contract into one document covering the whole process of delivering the project. It is the first standard form of contract for project partnering, it is the first multi-party building contract, and it is an architect-led initiative. It is claimed that merging a partnering agreement and a building contract could benefit the partnering process.

The result is a substantial document, with a closely interlocking set of terms, which is complex and not immediately easy to understand. Realistically the ACA advises that it should be used with the benefit of experienced legal and or other professional advice on its implementation.

Since it was launched by Sir John Egan in September 2000, it has received the recommendation of the Housing Forum, the Movement for Innovation, the Local Government Task Force, and the Construction Best Practice Programme. It has also been endorsed by the Construction Industry Council and the Housing Corporation.

Nature

Between the glossy covers, there are 54 pages covering the Project Partnering Agreement, the Partnering Terms, and Appendices.

The Partnering Agreement is signed or most likely executed as a deed, by the Client, the Constructor, Client's Representative and each consultant or specialist member of the Partnering Team. The Agreement will carry details of the Project, the site,

PPC2000

composition of the Partnering Team, Partnering Documents, and Core Group composition. The Design Team and Lead Designer are identified, and any amendments to the design development process as described in Section 8 of the Partnering Terms noted. Details are entered on other matters usually found in the Appendix or Contract Data with conventional contracts, but here also including matters such as incentives and insurance cover to be carried by each member of the Partnering Team.

The Partnering Terms are set out under 28 headings. The language is plain English, but some terminology is peculiar to this contract and may be unfamiliar. Helpfully there is a full set of definitions. There are five Appendices, the content of which may be summarised as follows:

Appendix 1: Definitions

Appendix 2: Form of Joining Agreement (with a Project of long duration inevitably there will be changes to the Partnering Team over time, and this is a mechanism for bringing in new joining parties who will be bound by the already established obligations)

Appendix 3:

- Part 1. Form of Pre-Possession Agreement (this is in essence an agreement to cover preliminary or enabling works to be undertaken by the Constructor)
- Part 2. Form of Commencement Agreement (confirmation by the Partnering Team that the Project is ready to proceed to commencement of work on site)

Appendix 4:

- Part 1. Insurance of Project and Site
- Part 2. Third Party Liability Insurance
- Part 3. Professional Indemnity or Public Liability Insurance
- Part 4. Insurance, General

Appendix 5:

- Part 1. Conciliation
- Part 2. Adjudication
- Part 3. Arbitration (if applicable)

PCC2000 is unique and crosses traditional boundaries. It is a combination of project management principles, legal conditions and procedural rules. It is logically structured, with commendable cross-referencing. It holds out the prospect of an integrated team approach and seamless delivery of the Project, but it calls for a high degree of commitment on the part of all concerned.

Use

Reports indicate that PCC2000 has been successfully used for both private and public sector projects ranging in value from multi-million pounds down to £600,000.

PPC2000

Partnering depends on an effective management structure, attentive administration, and good communications. The Partnering Team is the key to this and it will be beneficial to set it up as soon as possible. PCC2000 accepts that the composition is likely to change during the progress of the project. Ideally there should be opportunity to bring consultants, key specialists, Constructor and some sub-contractors and suppliers together at pre-construction stages. The Partnering Team members' liabilities are proportional to their responsibilities, as are incentive payments.

A Core Group is to be established by the Partnering Team Members, with responsibility to meet regularly to review and stimulate progress of the Project. Partnering Team Members must comply with decisions reached by the Core Group.

The Client's Representative has considerable authority. He or she may call, organise, attend and minute meetings of the Core Group and Partnering Team, and may issue instructions to the Constructor as empowered by the Partnering Terms. He or she will also be responsible for organising partnering workshops for the Partnering Team. However, restrictions can be placed on his or her authority and these are to be entered in the Project Partnering Agreement.

The Partnering Adviser will be a person who brings enthusiasm and a knowledge of partnering, and preferably already has a good track record. He or she has a very wide remit which includes reviewing all contracts for consistency with the Partnering Documents; preparing any Partnering Charter; preparing any of the agreements listed in Appendix 3; giving advice on the partnering process, partnering relationships, and partnering contracts; attending relevant meetings of the Core Group and Partnering Team; and assisting in the solving of problems and resolution of disputes. A tall order!

The Partnering Documents are listed as being:

- the Project Partnering Agreement;
- the Partnering Terms;
- the Partnering Timetable;
- the Commencement Agreement;
- any Partnering Charter;
- Consultant's Services Schedules and payment terms;
- the Project Brief;
- the Project Proposals;
- any Joining Agreements;
- the Price Framework;
- the Key Performance Indicators;
- any other Partnering Documents.

Unless there is anything to the contrary, this is the hierarchy of documents which prevail in the event of discrepancy or dispute.

PPC2000

The ACA advises that at the time of signing the Project Partnering Agreement, the team should have agreed the following:

- Client's Project Brief and the Constructor's Project Proposals;
- an initial Price Framework;
- provisional Key Performance Indicators;
- Consultant's Services Schedules and payment terms for those appointed by the Client.

With so many separate arrangements, this agreement cannot be stated in terms of a lump sum contract. The Client is responsible for payment to the consultants of agreed amounts properly due under the Consultant Payment Terms, and for payment to the Constructor of agreed amounts properly due in respect of Pre-Possession Agreement activities, and an Agreed Maximum Price calculated by reference to the Price Framework and other relevant Partnering Documents.

Synopsis of clauses

(Despite the fact that this is a unique type of contract, for the purposes of comparisons the same headings are used as those applied to more conventional contract forms earlier in this book.)

1 Intentions

- Roles, expertise and responsibilities are described in the Project Brief and Consultant's Services Schedules (Partnering Agreement) and Team Members work in a spirit of trust, fairness and mutual cooperation for the benefit of the Project (1·3).
- Partnering Objectives which apply to each member of the Partnering Team including Client and Constructor, are set out under six headings in clause 4·1 and cover design stages through to completion of the Project within the agreed time, price and quality. Also included are objectives such as innovation, improved efficiency, cost-effectiveness, lean production, reduction of waste, and measurable continuous improvement by reference to Key Performance Indicator targets (4·1).
- Partnering Objectives are followed by Partnering Targets under 10 headings, and each member of the Partnering Team undertakes to pursue these for the benefit of the Project and for the mutual benefit of the Team Members (4·2).
- In all matters the Partnering Team Members shall act reasonably and without delay (1·7).
- The Partnering Documents govern the relationships between Partnering Team Members (2·1).
- The Partnering Documents comprise the Partnering Agreement, Partnering Terms, together with any of the documents listed in clause 2.2.
- Priority of documents in the event of discrepancy is in descending order as listed in clause 2·6.

PPC2000

- Partnering Team Members work to achieve transparent and cooperative exchange of information, and integrate activities as a collaborative team (3·1).
- Communications between Team Members are to be in writing except where otherwise agreed (3·2).
- Team Members are to establish a Core Group, membership as listed in the Project Partnering Agreement (3·3).
- Decisions of the Core Group are by consensus, and Partnering Team Members must comply with authorised decisions (3·6).
- Partnering Team Members operate an early warning system, and each member notifies others as soon as he or she is aware of matters adversely affecting the project (3·7).
- Meetings of the Partnering Team are convened by the Client's Representative as scheduled or requested, and will normally be chaired by the Client's Representative. Only matters on the agenda are dealt with, and decisions are by consensus (3·8).
- Partnering Team Members are to develop arrangements for secondments, office sharing arrangements, access to computer networks and databases etc. as may benefit the Project (3·10).
- Partnering Team Members pursue together joint initiatives which might benefit the Project and such initiatives are considered by the Core Group (24·1).
- Partnering Team Members shall keep records as required by the Partnering Documents and permit inspection by other members of the Partnering Team (3·11).
- The Client's Representative is to act in accordance with the Partnering Terms and other Partnering Documents to facilitate an integrated design, supply, and construction process (5·1).
- The Client's Representative is authorised to represent the client in all matters, except membership of the Core Group, and always subject to any restrictions stated in the Project Partnering Agreement (5·2).
- The Client's Representative may issue empowered instructions to the Constructor (5·3).
- The Client's Representative is to call, organise, attend and minute meetings of the Core Group and Partnering Team Members as required or scheduled (5·1).
- The Client's Representative organises workshops for the Partnering Team Members, and organises and monitors contributions of Partnering Team Members to value engineering, value management and risk management (5·1).
- The Partnering Adviser as named in the Project Partnering Agreement may be replaced at any time by a decision of the Core Group (5·7).
- The Partnering Team Members may seek the advice and support of the Partnering

PPC2000

Adviser on a range of matters, including those listed in clause 5·6.

2 Time

- The Partnering Timetable is a Partnering Document (2·6) and covers the activities of the Partnering Team Members during the pre-construction period.
- The Project Timetable covers the period of construction following the Commencement Agreement.
- Members of the Partnering Team are to proceed regularly and diligently in the stages and by the dates in the Partnering Timetable (6·1).
- The Project Timetable is to be annexed to the Form of Commencement Agreement, and entries will show the date of possession and date for completion (6·2). The Constructor will submit the proposed timetable to the Client's Representative for review by the Core Group and approval by the Client (6·2).
- Where the Project is to be completed by sections, then completion dates will relate to each section and the Project (6·3).
- Possession of the site by the Constructor may be exclusive or non-exclusive, and programming may take this and any arrangements for deferred possession and interrupted possession into account (6·4).
- The Client's Representative may instruct acceleration, postponement or resequencing of any date or period in the Project Timetable (6·6).
- The Constructor will update the Project Timetable regularly and circulate it to the other Partnering Team Members (6·7).
- The Constructor is to use best endeavours at all times to minimise any delay or increased costs in the Project (18·3).
- An appropriate extension of the date for completion may be given for any one of 16 reasons listed in detail, all due to matters beyond the Constructor's control and including some neutral causes (18·3).
- The Constructor must notify the Client's Representative as soon as he becomes aware of any of the events listed, and supply appropriate evidence and detailed proposals for overcoming the events or minimising their impact (18·4).
- The Client's Representative must respond within 20 working days of the notification and make a fair and reasonable extension of time. The Client or the Constructor has 20 working days from the date of the Client's Representative's notice to dispute the award (18·4).
- An extension of time which affects consultants, and is not caused by their default, will bring an equivalent extension of time for performance of Consultant's Services (18·7).

PPC2000

- The Constructor will give the Client's Representative five working days' notice when he considers that Project Completion has been achieved. The Client's Representative is invited to inspect and test as appropriate (21·1).
- The Client's Representative, together with other appropriate Partnering Team Members, shall inspect and test, and within two working days following completion of this, the Client's Representative shall issue a notice to the client and Constructor either confirming that Project Completion has been achieved, or indicating aspects of the Project which the Constructor must rectify (21·2).
- Following completion of the Project the Constructor must rectify any defects, excessive shrinkages or other faults in the Project, which are due to materials, goods, equipment or workmanship not in accordance with the Partnering Documents. The Defects Liability Period is to be entered in the Project Partnering Agreement (21·4).
- The Client's Representative shall issue a notice to the Client and the Constructor confirming that the defects have been rectified (21·5).

3 Control

- None of the rights or obligations of the Partnering Agreement may be assigned or sub-contracted without the prior consent of all the other Partnering Team Members (25·2).
- Specialists and Preferred Specialists may be included in the Partnering Team. The Constructor is responsible for the performance of the specialists, except for any appointed direct by the Client (10·12).
- Instructions to the Constructor are given by the Client's Representative, and in accordance with the methods of communication for the Partnering Team (3·2).
- Instructions may require opening up for inspection or testing of any part of the Project, and rectification at no cost to the Client of any designs, works, services, materials, goods or equipment that are defective or otherwise not in accordance with the Partnering Documents (5·3).
- The Constructor can raise objections to an instruction for specific reasons, within two working days of issue of the instruction (5·4).
- The Constructor must promptly carry out empowered instructions issued by the Client's Representative. If it fails to do so after five working days of a further notice from the Client, the Client may pay another party to carry out the instruction and the cost shall be borne by the Constructor (5·5).
- Any Partnering Team Member may propose a Change to the Client, and proposed Changes shall be considered by the Client and the Client's Representative, and if approved will be notified by the Client to the Constructor (17·1).

PPC2000

- The Constructor within 10 working days will then submit to the Client a Constructor's Change Submission setting out the likely effects in terms of cost and progress (17·2).
- Within five working days from the Constructor's Change Submission, the Client's Representative will either instruct the Constructor to proceed (subject to reservation of any aspects until later) or withdraw the Change (17·3).
- The Constructor and specialists are to use and supply materials, goods and equipment of the types and standards stated in the Partnering Documents (16·2).
- The Constructor is responsible for security of the Project and the site (15·3).
- Ownership of materials, goods and equipment pass to the Client when they are incorporated into the Project, or when the Constructor receives payment for them. Such unfixed materials must not be removed from the site, must be stored separately and clearly marked as owned by the Client (15·4).
- Partnering Team Members are to implement a Quality Management System as set out in the Project Brief, Project Proposals, and Consultant's Services Schedules (16·3).
- From the date of the Commencement Agreement until Completion Date, the Constructor is responsible for managing all risks associated with the Project and the site, unless otherwise agreed (18·2).

4 Money

- The Constructor is to be paid in accordance with the Partnering Terms and the Price Framework (Project Partnering Agreement). The amounts for pre-possession activities are as entered in the Form of Pre-possession Agreement and the Client undertakes to pay these (12·2).
- The Agreed Maximum Price will be developed by reference to the Price Framework and other Partnering Documents and is to be as entered in the Form of Commencement Agreement. This is the sum payable by the Client to the Constructor, subject to increases or decreases in accordance with the Partnering Terms.
- Any fluctuation provisions must be set out in the Price Framework and Consultant Payment Terms (20·10).
- If the Partnering Documents link payment to performance targets stated in the Key Performance Indicators, then when the level of achievement of the Constructor or each consultant is demonstrable, the Client's Representative will determine the consequential additional or reduced payment (13·5).
- Where an extension of the Completion Date is awarded for certain events, the Constructor shall be entitled to additional payment in respect of site overheads and unavoidable additional work or expenditure (18·5 and 18·6).

PPC2000

- Applications for payment by the Constructor and by each consultant is made to the Client at the intervals stated in the Project Brief, or at the end of each calendar month. Payment can also be related to payment milestones, activity schedules etc. as set out in the Price Framework. Applications must be accompanied by details as stated in the Project Brief, and such further information as the Client's Representative may reasonably require (20·2).
- The Client's Representative will issue a valuation within five days from receipt of the Constructor's application, specifying the amount of payment proposed to be made and the basis of calculation all in accordance with the Housing Grants, Construction and Regeneration Act 1996. Final date for payment by the Client is 15 working days from date of issue of valuation, or 10 working days from receipt of VAT invoice from the Constructor whichever is later (20·3).
- Applications for payment to consultants by the Client will be subject to a notice issued by the Client, specifying the proposed payment to be made, and the basis of calculation all in accordance with the Housing Grants, Construction and Regeneration Act 1996. Final date for payment by the Client is 15 working days from the date of issue of the notice, or 10 working days from receipt of VAT invoice from the consultant whichever is later (20·4).
- Notice of withholding or deduction by the Client must be made not later than two working days before the final date for payment (20·6).
- Delay in payment by the Client will result in interest at the percentage specified in the Project Partnering Agreement (20·9) and may give rise to a right by a consultant or the Constructor to suspend performance until payment is received in full (20·17).
- Within 20 working days following Project Completion (or as stated in the Price Framework) the Client's Representative shall issue to the Client and the Constructor an account confirming the balance of the Agreed Maximum Price due, and the Client and Constructor shall seek to agree taking into account any adjustments, and subject to the amount stated as retention in the Price Framework (20·14).
- Within 20 days from the date of issue of that account the Client's Representative shall issue a valuation for the agreed amount, or if not agreed the amount that the Client's Representative considers fair and reasonable. Final payment is due within 15 working days from date of issue of the valuation (20·14).
- Within 20 working days following notice by the Client's Representative confirming that the Constructor has fulfilled all obligations in respect of rectifying defects, the Client's Representative shall issue to the Client and the Constructor a Final Account. When agreed this will be conclusive evidence as to the balance of the Agreed Maximum Price due, and the Client's Representative shall then issue a Final Account valuation (20·15).

PPC2000

- If agreement on the Final Account is not reached within 40 working days from the date of issue either the Client or the Constructor may seek remedies for resolving the dispute as set out in clause 27 (20·16).

5 Statutory obligations

- Partnering Team Members must comply with all laws and regulations currently in force in the country stated in the Partnering Agreement (i.e. the applicable law) and in the country in which the site is located, and with all statutory and other legal requirements (25·4).
- The Constructor will act as Principal Contractor for the purposes of the CDM Regulations, and the Planning Supervisor will be the person named in the Project Partnering Agreement (7.1).
- All Partnering Team Members must fulfil their obligations under the CDM Regulations including development of the Health and Safety Plan (7·1). Although not expressly stated, this will also of course relate to the Health and Safety File.
- Each Partnering Team Member shall use reasonable skill and care to ensure that all individuals for whom he is responsible adhere to the Partnering Contract, and each Member will be liable to the other Team Members for any loss, damage, injury or death caused by employees under their control (7·4).

6 Insurance

- Insurance of the Project and of the site, including structures on it, will be the responsibility of the Constructor or the Client as shown in the Commencement Agreement in the joint names of the parties and with waivers of subrogation (19·1). Where stated in the Commencement Agreement, the Constructor is to take out insurance in respect of damage to property other than the Project, not caused by default of the Constructor or specialist or consultant and which could not reasonably have been foreseen.
- The risks to be insured against are those set out in Appendix 4 Part 1.
- Each Partnering Team Member is to take out and maintain third party liability insurance for the amount stated in the Project Partnering Agreement and in accordance with Appendix 4 Part 2 (19·3).
- Professional indemnity or product liability insurance is to be taken out by Partnering Team Members named in the Project Partnering Agreement in accordance with Appendix 4 Part 3 (19·.4).
- Further insurance as required by entries in the Commencement Agreement can include environmental risk insurance (19·5), latent defects insurance (19·6), and whole project insurance (19·7).

PPC2000

7 Termination

- The Client may terminate the appointments of all Partnering Team Members if he no longer wishes to proceed with the Project either because of failure to achieve the pre-conditions to a start on site as set out in clause 14.1 or for any other reason not foreseeable by the client prior to the date of the Commencement Agreement. Procedures for giving notice and the consequences are set out (26·1).
- The Client (or Constructor as appropriate) may terminate the appointment of a Partnering Team Member for material breach of the Partnering Contract (26·3).
- The appointment of a Partnering Team Member will automatically terminate in the event that the Member becomes bankrupt or insolvent (26·2).
- The Client may terminate the appointment of the Constructor for specified defaults or breaches of the Partnering Documents. Procedures for giving notice and the consequential actions are set out in the Partnering Terms (26·4).
- A Partnering Team Member may terminate his own appointment in the event of specified defaults or breaches of the Partnering Documents by the Client. Procedures for giving notice and the consequential actions are set out in the Partnering Terms (26·5).
- If after the date of possession it becomes impossible to proceed with or complete the Project due to specified reasons, the Constructor must give notice to the Client's Representative. A Core Group meeting must be convened to consider the position and possible solutions. If an acceptable solution cannot be found then the Client may suspend or abandon the Project (26·6).
- Termination of the appointment of any Partnering Team Member does not affect the mutual rights and obligations of that and the other Partnering Team Members (26·15).

8 Miscellaneous

- All Partnering Team Members are to use reasonable skill and care appropriate to their respective roles, expertise and responsibilities, and owe to each other such duty of care as stated in the Project Partnering Agreement (22·1).
- Each Partnering Team Member is to provide or obtain collateral warranties as listed in the Project Partnering Agreement (22·2).
- The Constructor is to obtain specialist warranties in favour of the Client (22·3).
- The Agreement can also include for a design obligation by the Constructor, and in this event the obligation under clause 22·1 can be amended in the Project Partnering Agreement whereby the Constructor accepts full responsibility to the Client for the design, supply, construction and completion of the Project including the selection and standards of all materials, goods, equipment and workmanship, and including any design undertaken before or after the date of the Commencement Agreement by any

other Partnering Team Member. The Constructor may also be required to warrant that the completed Project shall be fit for its intended purposes.

- Section 8 in the Partnering Terms otherwise states that design development is in the hands of the Lead Designer and other Design Team Members, who are to develop the design with the object of achieving best value for the Client (8·1).
- At pre-commencement stages the Lead Designer submits outline designs to the Client and Core Group, and following Client approval, developed design is submitted to the Client and Core Group with detail sufficient for a full planning application (8·3).
- Following Client approval, and after Core Group consultation, the Lead Designer applies for full planning permission, and with other Design Team Members brings the design to the level necessary for the selection of specialists, development of the Price Framework, and satisfying of planning conditions and other regulatory approvals (8·3).
- After commencement all further design work is prepared and submitted to the Client and other Partnering Team Members for approval or comment in accordance with periods stated in the Project Timetable (8·6).
- Each Partnering Team Member retains intellectual property rights in all designs and other documents that he prepares for the Project, but grants to the Client and other Partnering Team Members a licence to copy and use such designs relating to completion of the Project (9·2).
- Nothing in the Project Partnering Agreement or Partnering Terms confers any benefits or rights to third parties, unless expressly stated otherwise (22·4).
- Nothing in the Partnering Documents creates a partnership between Partnering Team Members (25·1).
- Any special terms to be imported into the contract must be identified as special terms by reference to this clause and must be set out in or attached to the Project Partnering Agreement or the Commencement Agreement (28).

9 Disputes

- In the event of any difference or dispute with other Partnering Team Members, a Member must give notice to the other Members and the Client's Representative (27·1).
- The Partnering Team Members involved are to apply the Problem-Solving Hierarchy shown in the Commencement Agreement, guided as necessary by the Partnering Adviser (27·2).
- Where use of the Hierarchy fails to provide an acceptable solution within a stated timetable, the Client's Representative will convene a meeting of the Core Group in an attempt to reach an agreed solution (27·3).

- If the dispute is still not resolved, the parties may chose to refer the matter to conciliation as described in Appendix 5 Part 1, or mediation, or any other form of alternative dispute resolution (27·4). The conciliator may be named in the Project Partnering Agreement.
- The parties involved may exercise the right to refer their difference to adjudication (27·5) in accordance with Appendix 5 Part 2. The adjudicator may be named in the Project Partnering Agreement.
- If the difference or dispute is not finally resolved by adjudication, the parties may refer the matter either to arbitration or to the courts (27·6). Arbitration is covered in Appendix 5 Part 3. The nominating body for an arbitrator may be named in the Project Partnering Agreement.

This contract?

If considering using PPC2000 remember that:

It is the only standard contract specifically drafted for project partnering.

The conditions are plainly worded and easy to read. However attractive the notion of a single contract bringing all important parties together in a binding relationship is, inevitably this calls for an open and receptive mind.

Contract administration and project management under it seem likely to prove demanding, but the contract has apparently been used in practice with successful results.

If the ultimate in partnering arrangements is desired, then this form, which is the result of an architect -led initiative, has no competitors as yet.

Related matters

Documents

ACA Standard Form of Contract for Project Partnering PPC2000
ACA Standard Form of Specialist Contract for Project Partnering SPC2000

References

Introduction and Explanatory Notes (included with PPC2000)

14 Choice scenarios

Taking into account

There will be times when operations cannot be tidily covered by a single standard form of contract. For example, some work may be needed ahead of the main contract, or the need for some additional specialist work may become apparent only as the main contract works proceed. This may bring the need for enabling contracts, or separate contracts proceeding in sequence or in parallel, and a combination of otherwise unrelated standard agreements. Care is obviously needed in such circumstances to make sure that there is no conflict of responsibilities, and that the rights and obligations of contracting parties are clearly set out.

Standard forms currently published will adequately cover most situations, although this might necessitate the use of option clauses or supplements as provided for in the particular contract form. There may be exceptional circumstances where only a bespoke agreement will give a satisfactory answer, and if so then this should be drafted by an experienced construction lawyer instructed directly by the client.

At all times the contract should try to take account of eventualities that can be foreseen, and to ensure that the intentions of the parties are expressed clearly, with certainty, and that the allocation of risks is as intended.

In recent years statute law has impinged on the common law of contract, and affected building contract conditions. It is necessary to take into account applicable legislation through expressly stated provisions, or at least to have an awareness of what might be implied. Legislation which might need to be considered includes:

Construction (Design and Management) Regulations 1994 (SI 1994/3140)

This will apply to nearly all temporary and permanent works. The statutory duties for the Employer and Contractor are usually made contractual obligations also.

Check whether: the contract terms take into account the role of the Employer in respect of appointments, and the obligations of the Planning Supervisor and Principal Contractor in respect of the Health and Safety Plan and File.

Unfair Terms in Consumer Contracts Regulations 1994 (SI 1994/3159)

This legislation applies to contracts for goods and services between a consumer and a supplier. The former will be a natural person acting in a personal way (e.g. a home owner) and the latter a person acting in the course of business (e.g. a consultant or a builder). This is primarily a consumer protection measure, and applies to any term in a contract which has not been individually negotiated. It calls for fair terms, to be expressed in plain intelligible language. An unfair term will not be binding on the consumer. In assessing the requirement of good faith, the bargaining strength of the parties will be taken into account.

Check whether: one of the parties is a consumer, and if so care is needed to ensure that the contract complies with the requirements for fair terms and plain language. Most standard forms have not been individually negotiated. Hence the present attempts to publish building contracts which are 'consumer contracts' for smaller domestic works. Care is needed if drafting special clauses in building or consultant appointment agreements to make certain that the terms are understandable and understood by the consumer.

Housing Grants, Construction and Regeneration Act 1996, Part II Construction Contracts

This will apply to 'construction contracts' and this definition will include contracts for professional services, interior or exterior decoration, landscape and building contracts. It will not apply to contracts with a residential occupier, provided that the work is principally on a dwelling for owner occupation.

Check whether: the contract terms expressly include the right to refer disputes to adjudication, and whether payment procedures meet the requirements stipulated in the Act. If the Act is applicable and yet not expressly included for, then the Scheme for Construction Contracts Regulations will automatically apply.

Party Wall etc. Act 1996

As with much legislation relating to development such as the Town and Country Planning Acts, Building Acts etc., this will not usually feature in the contract Conditions, although where party wall agreements or awards are concerned this might bring significant implications for contract administration.

Check whether: it is likely that work will affect party walls, because if so notices might be issued or received during the course of the works, and construction work might be affected due to a party wall award. Contract Conditions should provide for instructions to cover postponement, variations, extensions of time, disturbance costs, etc. which might become relevant.

Contracts (Rights of Third Parties) Act 1999

It used to be held that at common law only the parties to a contract had obligations and benefits from it. The agreement touched only the parties, and third parties were outside the contract. Now this piece of legislation has brought the right of a third party to enforce contractual terms, always provided that the contract expressly provides for this or purports to confer a benefit, and that the third party is identified in the contract. The contract may expressly exclude or limit liability.

Check whether: it is intended that certain third party rights are to apply, and if so, whether the contract expressly includes this in the manner required by the Act. If uncertain on this matter, it might be advisable to obtain legal advice. The majority of published standard forms of building contract now expressly state that the rights of third parties under this legislation will not apply.

Scenarios on choice decisions facing the architect

Preliminary or enabling works contracts

Scenario A

The tennis club in the affluent village of Fairview has a benefactor who has promised to donate a new prefabricated pavilion, produced by a subsidiary to the company he owns. This will be supplied and erected as a package. Internal decoration will be needed, which the members feel they can undertake themselves. External works will also be required, which must be completed prior to delivery. A local contractor will be needed to carry out this work which includes the access road, hard standing, bringing the site to proper levels, drainage, and the necessary concrete base for the pavilion.

Which contract?

According to the nature of the work for this preliminary contract, consider a lump sum contract. A minor works form such as JCT MW98 might be appropriate if the work is relatively straightforward.

Comment

Difficulties can arise over phasing where there are sequential contracts, particularly when different contractors are involved. The contractor responsible for erection must be able to rely on getting unimpeded possession on the due date and the site being ready to the agreed state. Practical completion of the preliminary contract will need to be certified, and the respective liabilities for defects and damage clearly established at the beginning (for example, MW98 only refers to defects in work under that contract, being made good by the original contractor). A separate Health and Safety Plan and File contribution will probably be required.

Scenario B

Planning permission for the latest 'Homeforce' DIY Superstore contained a condition that the main facade of the 19th-century Boon Mills, which now occupies the site, should be retained. Site clearance is imminent and the selected part of the listed facade needs to be stabilised and protected in what could be a delicate operation. It is decided that this work should be entrusted only to an expert demolition contractor.

Which contract?

A separate contract is required for the demolition work. Consider a suitable minor works building contract e.g. GC/Works/2 (1998) or, better still, a specialist contract (e.g. the Standard Form of the National Federation of Demolition Contractors). Make sure that the Specification and Conditions include relevant provisions for indemnity and insurance, and that the contract for the building works allows for another contractor to work on the site occupied by the main contractor.

Comment

The demolition contractor needs to be given clear information about respective obligations for the safety and protection of the site. If material is to be salvaged and stored, it should be clearly stated. (The employer might expect to be credited if the contractor is allowed to acquire salvaged material.) Demolition work can be dealt with as a preliminary contract or as a sub-contract of the main contract. If the former, then a Health and Safety Plan specific to this work might be required or it might be contained in the Plan produced by the principal contractor.

Scenario C

Victoria Towers has been allowed to deteriorate ever since a disastrous fire last century. Now Country Heritage is prepared to fund substantial restoration and has commissioned a detailed survey and report on the condition of the fabric. A contractor will be needed to carry out clearance and opening-up to allow investigation work, the precise extent of which cannot be known at the outset.

Which contract?

Consider contacting a reliable builder, preferably one with experience of this kind of work who employs and personally supervises craftsmen with a knowledge of the relevant materials and methods. This is unlikely to be a lump sum contract, and a cost plus approach may be the only practicable one. It is doubtful whether any of the standard building contracts effectively covers this kind of operation, and it is probably best dealt with on the basis of an exchange of letters and an agreed Schedule of rates, costs and profit.

Comment

Proper protection and safety measures are the contractor's responsibility, but check that these are not skimped. Such work needs close direction, and the contractor should appreciate when quoting rates that this might not be a single continuous operation. Returns to site for further investigation are a likely requirement.

Trades contracts for work of an intermittent nature

Scenario D

Champers is a popular cellar bar and restaurant in Westville. Success has brought more sophisticated patrons, a need for expansion, and a more chic ambience. Further vaults have recently become available, and these development ideas can now be realised. The restaurant's reputation is such that there must be no complete shut-down during building work, which may need to be carried out in a rather piecemeal fashion. What is equally important, patrons of Champers must at all times be able to enjoy their food and wine safe from any intrusion of dirt, dust, unsavoury smells and unwelcome noise resulting from the renovations.

It has been suggested that whilst builders' work and attendance can be provided by one reasonable building contractor, services installations and bar and kitchen work must be left to specialists. It is envisaged that the work will be carried out in a periodic or phased manner possibly over an extended period, and that integration and overall direction will be critical. The architect agreed to act as project manager, but the situation now suggests that the role is developing more into management of separate trades contracts.

Which contract?

The situation calls for sound management, effective coordination, and firm control. Because of the intermittent nature of the work, a standard lump sum contract with the builder is thought to be unlikely to achieve the desired result. It is thought that this could best be carried out by a consultant operating as a management contractor. However, the capital costs involved are likely to be relatively small, and probably separate trades contracts entered into direct between the client and each specialist firm would be the most satisfactory answer. A short form (e.g. JCT MW98, one of the ASI forms, or GC/Works/4) might be an appropriate document to use for these, although there might also be a need to include additional conditions.

Comment

Any agreement, on whatever form, should include for matters such as the following:

- Time: state dates for commencement and completion. Consider problems of phasing and possession.
- Money: state basis for valuing work done. State when payments are to be made and the procedures involved. State the amount of retention.
- Control: state the need for architect's instruction before any deviation is made. Establish procedures to ensure integration and coordination of the various trades involved, who is to be responsible for setting up site access, welfare provisions and storage facilities etc.
- Insurance: state who will be responsible for insuring the existing structure, contents and new work. Consider cover for any consequential losses arising from the carrying out of the work.
- Termination: if this is to be included, state whether it is an option open to either party and if so on what grounds. Trades contracts are almost certain to be construction contracts to which adjudication would be a statutory requirement.

It might also be worth considering whether there should be a provision which, in the event of key contractors falling behind, would allow the architect to take action (at no extra cost to the employer) to bring the work back to time. This might be by bringing in an additional labour force, or even another firm.

An architect involved in such a management role would need an appropriate contract for professional services with the employer, and should expect an appropriate fee.

Contracts with substantial specialist content

Scenario E

The catalytic degrader at Hotwells Heavy Water and Associated Products (1981) Ltd has become redundant, and is due to be replaced by the very latest installation from Superlink Fibreoptics. The new detached plant room will be a relatively simple structure, but the services installation it is to house is of mind-blowing complexity and far exceeds the capital cost of the actual building work.

Which contract?

If the entire operation is to be covered by one building contract, then consider a lump sum contract (e.g. JCT98) which allows for the installation specialists to be nominated sub-contractors.

Alternatively, the specialist installation firm could be considered as the principal contractor with overall responsibility for carrying out the work. The builder would then be a sub-contractor responsible for carrying out work on the building envelope including necessary attendant builders' work. Consider either a lump sum building contract (e.g. JCT98) or an appropriate engineering contract as the main form. Where appropriate, and if the work is conveniently self-contained, there could be separate parallel contracts for the building work and the engineering work.

Comment

The former NJCC Procedure Note 4 Placing of Contracts with a Substantial Building Services Engineering Content, advised that where parallel contracts are used, great care is needed to ensure effective coordination and to eliminate the risk of duplication. There might also be complications where consultants with differing functions and duties under different contracts of engagement are employed on the same project.

Contracts for landscape work

Scenario F

For the proposed Eventide Homes cluster development, specially designed with the needs of the over-60s in mind, it is thought that the external landscaping needs to provide a particularly tranquil setting. The architect-led design team includes a landscape architect, and although both soft and hard landscaping are to be carried out, it is thought best to treat this as a sub-contract to the main building contract. It is intended to use a particular landscape firm as sub-contractor.

Which contract?

If the landscape firm is to be named under IFC98 or nominated under JCT98, then the appropriate forms either NAM or NSC must be used. They might not prove suitable.

Comment

The Specification or bills of quantities preambles should make specific reference to special provisions to cover such matters as plant failure and maintenance, and malicious damage or theft before practical completion. Such items would not ordinarily appear in a building contract.

Scenario G

Developers Rushe & Roulette have acquired a redundant office block in the City for the proverbial song, and promptly set about converting it to make apartments attractive to young executives. The site needs the attentions of an innovative landscape architect, and much of this will be hard landscape using non-traditional materials and specialist technology. It is therefore proposed that although this work will proceed at the same time as the building work it should be undertaken as a quite separate contract.

Which contract?

Consider a lump sum contract (e.g. JCT IFC98) perhaps modified after taking account of relevant JCLI Practice Notes. Alternatively, if the type of work suggests that remeasurement is more practicable (e.g. if there is a substantial amount of earth-moving, contouring and associated work of a civil engineering type), it may be more appropriate to consider the ICE Conditions of Contract for Minor Works.

Comment

If this is a self-contained contract for landscape work, it is probably best administered by a landscape architect. Careful coordination will be needed if it is a contract in parallel with another building contract.

Scenario H

The new council offices under construction at Tan-y-groes have attracted a great deal of media interest. Work has not proceeded to programme and failure to complete by the contract completion date would be a matter of intense civic embarrassment, as arrangements are already in place for an opening ceremony by a distinguished person. To make sure that this will go ahead as planned, it has been decided to omit all the landscape work to the central sculpture court. The building can be occupied and the landscape work can proceed at a later date under a separate contract.

Which contract?

Consider the JCLI Form of Agreement for Landscape Works. This is closely modelled on the JCT MW98 contract, but includes additional provisions for dealing with situations such as plant failure, malicious damage or theft, after-care and maintenance. The latter will require a separate JCLI Agreement.

Comment

It is perfectly acceptable for the administrator of the JCLI landscape contract to be other than a landscape architect (e.g. an architect), but the duties presuppose a knowledge which an architect experienced only in building operations might not possess.

Contracts for fitting out

Scenario J

Ground floor shop units on the new Castle Gates Development are to be left as shells, to be fitted out by tenants. They will assume responsibility for fitting out the interiors and for providing shop fronts, under direct contracts quite separate from the developer's contract with the principal contractor.

Which contract?

Shop-fitting and interior work could be carried out under a lump sum contract in respect of each unit. At its simplest this could be by written acceptance of the specialist firm's offer. However, such quotations are often subject to each firm's own conditions, and these might not be acceptable. Architects acting for tenants might be wiser to suggest an appropriate standard form of contract (e.g. JCT MW98 or IFC98, depending on the scale and nature of the work). Alternatively an agreement suitable for interior work is available from the Chartered Society of Designers (formerly the Society of Industrial Artists and Designers).

Comment

If the main contractor is still on site, care should be taken to establish respective site responsibilities and in particular proper insurance arrangements. The principal contractor might require other contractors to work in accordance with the Health and Safety Plan for the site, or if the tenant's work is carried out under separate unrelated contracts, then a Health and Safety Plan for these works might be necessary.

Joint venture contracts

Scenario K

The oil rich port of Fyl-yr-Up requires new state of the art terminal facilities. This is likely to be a large-scale and sophisticated construction operation calling for significant engineering installations which must be coordinated and integrated with great precision into a series of buildings which will operate as the necessary plant housings. The conventional arrangement of appointing a main contractor and various sub-contractors specialists is not likely to prove satisfactory. A tremendous amount of detailed planning and coordination will be required before the complex work starts on site, and there must be absolutely no risk of delays or disruption because of installation or integration problems.

Which contract?

Consider a joint venture approach, partnering or a management contract. If joint venturing, then the selected tendering companies must be willing to combine under one legally constituted partnership and accept joint and several liability. The use of a traditional lump sum form (e.g. JCT98) would not necessarily be precluded, as the contractor would be a single entity, albeit a specially formed partnership or company.

Comment

The partnership agreement should provide for joint and several liability and a performance bond would normally be required. This method of working can be particularly effective where design input is required, with elements of the design contributed by the various specialist partners, and all effectively coordinated within the larger organisation.

Completion contracts

Scenario L

The contractor for a new group practice veterinary surgery in Fairmeadow became insolvent after the seventh week of a 36-week contract. His employment was automatically determined under the terms of the contract, and it is accepted that reinstatement is out of the question.

Which contract?

Select a new contractor (ideally under the same contract Conditions as the original) to take the place of the first contractor. This might be achieved through negotiations with the second lowest tenderer on the original tender list, but if the work has been made more onerous by the efforts and departure of the failed contractor, there will certainly be an increase in the contract sum.

Comment

If the circumstances or climate for tendering has changed markedly since the original tenders, it might be necessary to amend the original documents and invite new competitive tenders. An insolvency practitioner acting on behalf of the original contractor should be consulted and must be kept fully informed of all developments. Employer's costs and expenses arising out of the determination might be part of the claim to be brought against the original contractor.

Scenario M

Work on the Limboland Fitness Centre was well over two-thirds complete before the contractor's performance became progressively slower and somewhat erratic. It was no great surprise when a letter arrived from a firm appointed to act as receiver. After discussions, it was generally agreed that the most satisfactory course of action is to have the contract completed by a substitute contractor.

Which contract?

The contract just might be completed by reinstatement and assignment. Here the substitute contractor accepts responsibility for completing under the terms of the original contract. A deed of assignment would be required as agreed between the employer, receiver acting on behalf of original contractor, and the substitute contractor.

Alternatively, the contract might be completed by reinstatement and novation. In this case, although the substitute contractor might accept responsibility for completing the works, all the terms of the original contract might not be acceptable. A deed of novation would be required as agreed between the employer, receiver acting on behalf of the original contractor, and the substitute contractor.

Comment

Where a substitute contractor is willing to complete the work, an adjustment of the completion date is almost inevitable. The contractor might also have reservations about accepting responsibility for work already carried out and perhaps covered up, and additional work might be inevitable. The risks of taking on work undertaken by a previous contractor can often be squared by an additional single fixed premium payable to the new contractor. A new Health and Safety Plan might be required.

Scenario N

The Dragon Housing Association refurbishment scheme was complete except for a few outstanding items. With great reluctance, the architect yielded to pressure and certified practical completion to allow unhappy decanted tenants to resume occupation. Unfortunately, all efforts to bring back the contractor to complete the outstanding work failed. By the expiry of the defects liability period the contractor had become insolvent. Some money was held back at practical completion, in addition to half the retention figure. The issue now is to have the work completed.

Which contract?

The outstanding work could be carried out on a lump sum or preferably a cost plus basis by another contractor. An exchange of letters might be considered sufficient, but this would probably still be a 'construction contract' to which the payment procedures and adjudication option would apply. Depending on the extent and nature of the outstanding work, it might be preferable to use a short form (e.g. JCT MW98 or an ASI form).

Comment

A Specification or Schedule will be needed if competitive quotations are to be sought. The best course of action should first be discussed with the receiver or liquidator, who should be kept fully informed of developments.

Contracts for jobbing repairs or maintenance

Scenario O

When the date for judging this year's Best Kept Municipal Vista competition was announced, it was decided to undertake a clean-up and repainting of several High Street facades as a group operation. Rendering to the facades is in need of some renovation and some stonework requires attention. Woodwork and ironwork also needs repainting, and fascia boards relettered. This is mostly routine maintenance work, and well within the competence of the local builders.

Which contract?

For small-scale straightforward jobs, tenders might be sought on the basis of minimal drawn information, and a Specification or Schedule. This might result in just a lump sum, or if sufficiently itemised, the contractor's figures could give a useful breakdown. An exchange of letters might be sufficient in many cases, but where a larger amount of work is involved, then a standard form of contract should be used (e.g. JCT MW98 or an ASI form).

Comment

If the entire work is to be treated as one contract, then the client would need to be identified, and would presumably have authority to appoint a contract administrator. Channels of communication would need to be clearly established to avoid the risk of individual property owners issuing instructions directly. If the individual property owners prefer to enter into separate contracts with a common contractor, then assuming that the same standard forms are used, and the same contract administrator is appointed under each, coordination might be more exacting.

Scenario P

The New Cambrian Bank, which has premises in major towns throughout the Principality, is about to embark on a programme of regular maintenance and repair. It is envisaged that parcels of work on a regional basis will be offered for tender, in some cases with the added proviso that appointed contractors must also make themselves available to tackle emergency repairs at short notice.

Which contract?

Consider a term contract. Contractors can be selected by competition on a Schedule of rates. There may also be a standard call-out charge for emergency visits. The contract might cover a specified number of properties and run for a fixed time period. Orders would be issued to authorise ad hoc work. It is usual to advise the contractor of the approximate total value of the work envisaged, but any firm guarantees should be avoided.

Comment

The JCT publishes the Standard Form of Measured Term Contract MTC/98. Some bodies have their own model documents for such contracts. These contracts usually require the presence of a contract administrator appointed by the employer.

Scenario Q

The architect engaged by the Allday family has given them advice on what is needed to update the cottage they have just acquired. It amounts to a small amount of alteration work and some maintenance and repair work. It is agreed that this work can safely be entrusted to the excellent local builder.

Which contract?

Assuming that the architect's services are no longer required, and that this is a private residence for occupation by the Allday family, then a consumer contract (e.g. JCT Building Contract for a Home Owner/Occupier) would be a safe recommendation.

Comment

An exchange of letters, or a very basic formal agreement (e.g. JCT Contract for Home Repairs and Maintenance) could be considered as an alternative, but neither is likely to provide adequate contract Conditions. The fact that this is work to a private residence for owner occupation means that statutory requirements relating to adjudication and payment provisions under the Housing Grants, Construction and Regeneration Act 1996 will not apply.

Bibliography

Chappell, D.
Parris's Standard Form of Building Contract
3rd edn, Blackwell Science (2002)

Chappell, D. and Powell-Smith, V.
The JCT Minor Works Form of Contract
2nd edn, Blackwell Science (1998)

Chappell, D. and Powell-Smith, V.
The JCT Intermediate Form of Contract
2nd edn, Blackwell Science (1999)

Chappell, D. and Powell-Smith, V.
The JCT Design and Build Contract
2nd edn, Blackwell Science (1999)

Eggleston, B.
The New Engineering Contract
Blackwell Science (2000)

Eggleston, B.
The ICE Conditions of Contract, Seventh Edition
2nd edn, Blackwell Science (2001)

Jones, N.F. and Baylis, S.E.
Jones & Bergman's JCT Intermediate Form of Contract
3rd edn, Blackwell Science (1999)

Lupton, S
Guide to JCT98
RIBA Publications (1999)

Lupton, S.
Guide to MW98
RIBA Publications (1999)

Lupton, S.
Guide to IFC98
RIBA Publications (2001)

Lupton, S.
Guide to WCD98
RIBA Enterprises (2002)

Bibliography

Lupton, S. (ed)
Architect's Job Book
7th edn, RIBA Publications (2000)

MacRoberts, Solicitors
MacRoberts on Scottish Building Contracts
Blackwell Science (1999)